Monographs in Electrical and Electronic Engineering

Editors:P. HAMMOND, D. WALSH

R.L. Bell: *Negative electron affinity devices* (1973)
A.A. Bergh and P.J. Dean: *Light-emitting diodes* (1975)
G. Gibbons: *Avalanche-diode microwave oscillators* (1973)
F.N.H. Robinson: *Noise and fluctuations* (1974)
Richard L. Stoll: *The analysis of eddy currents* (1974)
H.C. Wright: *Infrared techniques* (1973)

The Gunn effect

G. S. HOBSON

CLARENDON PRESS · OXFORD
1974

Oxford University Press, Ely House, London W. 1

GLASGOW NEW YORK TORONTO MELBOURNE WELLINGTON
CAPE TOWN IBADAN NAIROBI DAR ES SALAAM LUSAKA ADDIS ABABA
DELHI BOMBAY CALCUTTA MADRAS KARACHI LAHORE DACCA
KUALA LUMPUR SINGAPORE HONG KONG TOKYO

ISBN 0 19 859318 X

© OXFORD UNIVERSITY PRESS 1974

Typeset by E.W.C. Wilkins Ltd., London
and printed in Great Britain
by J.W. Arrowsmith Ltd., Bristol

Preface

Gunn diodes and other derivatives of the transferred-electron effect are taking their places in microwave technology as well as creating new applications. This book caters for the need to include an understanding of transferred-electron devices in the skills of the electronic engineer at Master's degree and final-year undergraduate level. Throughout the emphasis is on simplified descriptions of the physical principles of the device and circuit mounting. Most of the ideas where clarified by or arose from discussions with numerous fellow workers who are recognized in the references. In particular, I would like to thank Professor P.N. Robson of the University of Sheffield for many lengthy discussions on all matters and the several postgraduates whom it has been my pleasure to supervise and interact with during investigations of Gunn-effect devices. On another plane considerable gratitude is owing to Pauline, Philip, and Julia.

Sheffield
23 *October* 1973

G.S. Hobson

Contents

Contents

<table>
<tr><td>

1

</td><td>

Introduction

</td></tr>
</table>

While J.B. Gunn (1963, 1964) was studying the noise properties of gallium arsenide under conditions of high electric field he observed that coherent oscillations occured in some devices when the field exceeded a critical value. Subsequently, Gunn observed similar effects in indium phosphide. This spontaneous discovery founded the development of active-semiconductor devices to replace microwave vacuum tubes, with advantages previewed in the development of the transistor.

For many years prior to Gunn's observations, proposals had been made on the possibility of obtaining negative resistance in the bulk of a semiconductor device without the intervention of a p—n junction or similar interface. One proposal, given independently by Ridley and Watkins (1961) and Hilsum (1962), contained the essential details of the explanation of Gunn's observations in the so-called transferred-electron effect. Transferred-electron devices utilize a relationship between the electron-drift velocity and the electric field of the form illustrated by the solid line in Fig. 2.1. The drift velocity increases ohmically with the electric field from the origin but passes through a maximum (the peak velocity v_p) at the threshold field E_T. There is a region of negative-differential mobility beyond E_T where the drift velocity falls asymptotically towards the valley velocity v_v. This behaviour is a consequence of particular properties of the conduction-band structure and electron-transport properties in gallium arsenide and some other similar materials (see Chapter 2).

Recognition that Gunn's observations could be explained by the transferred-electron theory was finally made by Kroemer in 1964. This linkage was obscured by (Bulman, Hobson, and Taylor 1972) (1) the multitude of freak oscillations that had been reported in the preceeding era, (2) the several theories that were contenders for explaining this new phenomenon which, of necessity, occurred under conditions of incomplete documentation, and (3) the complexity of the internal instabilities that occur in a medium exhibiting a bulk negative-differential mobility (see Chapter 2).

The various static and dynamic non-uniformities hinted at above are profoundly effected by the boundary conditions imposed by any real device (Chapter 3). Once we understand the effect of boundaries it is possible to exploit the advantageous features of transferred-electron devices and cope with their disadvantages in the developement of oscillators (Chapter 4) or amplifiers

(Chapter 5). As may be expected, the development of a new type of device generates it's own problems of fabrication (Chapter 6) and in the processing or transfer of information (Chapters 8 and 9).

An attempt has been made to give simple explanations of transferred-electron phenomena. The references given throughout are not intended to form necessary reading but rather to guide the interested reader to greater depth and detail. Also for this purpose the reader is referred to several scientific journals that have devoted special issues to microwave semiconductor devices. References are also made to several review articles for simple and readable introductions to the transferred-electron effect.

<table>
<tr><td>

2

</td><td>

The transferred-electron effect and space-charge instabilities

</td></tr>
</table>

2.1. The transferred-electron effect

The negative-differential mobility of electrons in GaAs (Fig. 2.1) arises from the particular form of the band structure which is shown in Fig. 2.2. At low electric fields, the conduction electrons occupy the bottom of the central valley and are distributed over the thermal energy range ($\sim$0·025 eV at 300 K). Finite

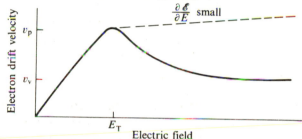

Fig. 2.1. The velocity–field characteristic of electrons in GaAs (solid line). The dashed line is a characteristic that may be obtained if $\partial\mathscr{E}/\partial E$ is small.

electric fields deliver energy to the electrons during their acceleration until they collide with imperfections of the crystal lattice. In the central valley of high-purity GaAs these imperfections are predominantly thermal lattice vibrations with a polar-optical character (Bulman *et al.* 1972). The collisions occur between the electrostatic fields of the electron and the dipolar charge of the crystal lattice. The latter is revealed by lattice-vibrational movement of the ionically charged gallium atoms and oppositely charged arsenic atoms from their equilibrium positions. On collision, the electrons lose a component of their momentum which is directed along the electric field. They also lose some kinetic energy which appears as Joule heating of the crystal lattice. The Coulombic nature of the polar-scattering mechanism largely changes the direction of the electrons' motion with a small change of its speed. Accordingly, the rate of momentum redistribution on collision is somewhat greater than the rate of kinetic-energy redistribution. Some of the kinetic energy gained from the application of an electric field is effectively shared between all the electrons when their momentum is redirected so that their mean energy rises. An increase in collision rate follows this increase of kinetic energy until equality again exists between the

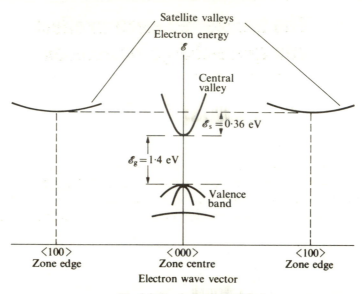

Fig. 2.2. Band structure of GaAs.

rate of energy delivery by the electric field to the electron and the rate of energy delivery by electrons to the lattice. As the electric field is increased further, the equilibrium state exists at a higher mean energy and the electrons occupy a broader range of energies in the central valley of the conduction band.

When the kinetic energy of the electrons exceeds 0·36 eV they have the additional choice of occupying the three satellite valleys of the conduction band which lie along $\langle 100 \rangle$ crystallographic directions. In these valleys, electrons have an effective mass of $\sim 0 \cdot 4 m_0$ which is approximately six times the central-valley effective mass ($0 \cdot 068 m_0$). Accordingly, the density of available electron states in unit energy interval in the satellite valleys is much greater than in the central valley (McKelvey 1966; Bulman *et al.* 1972). Those electrons which have the energetic capability will exist predominantly in the satellite valleys. As the electric field is further increased more electrons have sufficient energy for intervalley transfer so more of them occupy the satellite valleys. The electron mobility in the satellite valleys μ_2 is smaller than that in the central valley μ_1 by a factor of ~ 70, owing to the higher effective mass of the electrons and the stronger scattering processes (Bulman *et al.* 1972) which are operative in the satellite valleys. Even though an incremental increase of electric field will cause a separate increase of drift velocity in each valley, there is the possibility of a combined negative-differential mobility occuring by electron transfer. If the electron density in the central valley is n_1 and in the satellite valleys is n_2 the mean drift velocity $v(E)$ is

$$v(E) = \frac{(n_1 \, \mu_1 + n_2 \, \mu_2)E}{n} \qquad (2.1)$$

4

where $n(= n_1 + n_2)$ is the total electron density. Non-linearity of the mobility in each valley will be neglected for the present purposes. From eqn (2.1),

$$\frac{\partial \{v(E)\}}{\partial E} = \frac{n_1(\mu_1 - \mu_2) + n\mu_2 + (\partial n_1/\partial E)(\mu_1 - \mu_2)E}{n}. \qquad (2.2)$$

In order for $\partial \{v(E)\}/\partial E$ to be negative,

$$\frac{\partial n_1}{\partial E} < -\frac{n_1(\mu_1 - \mu_2) + n\mu_2}{(\mu_1 - \mu_2)E}. \qquad (2.3)$$

If E is not so large that intervalley transfer is almost complete n_1 is of the same order as n but $\mu_1 \gg \mu_2$, so that inequality (2.3) becomes

$$\frac{\partial n_1}{\partial E} < -\frac{n_1}{E}. \qquad (2.4)$$

Inequality (2.4) can only be satisfied if the electron energy in the central valley increases sensitively with the electric field. Otherwise a velocity-field characteristic similar to the dashed line in Fig. 2.1 will be obtained, and there will not be a negative-differential mobility. The possibility of a negative-differential mobility is made reality in GaAs by the nature of the polar-scattering mechanism. In a similar manner to the behaviour of Rutherford scattering of an electron by a fixed point charge, this Coulombic scattering mechanism becomes weaker as the electron kinetic energy increases (Bulman *et al.* 1972). At a critical electric field (peculiar to the material) the polar scattering can no longer transfer kinetic energy from the electrons to the crystal lattice as rapidly as it is supplied by the electric field. The electron energy 'runs away' under this condition (Fig. 2.3) of

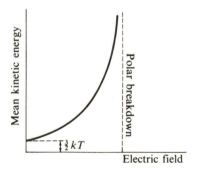

Fig. 2.3. The form of the relationship between the mean kinetic energy of electrons and the electric field in a material exhibiting 'polar breakdown'.

'polar breakdown' and causes dielectric breakdown in some materials. For GaAs the breakdown field is approximately 500 V mm^{-1}. Most of the increase of electron energy occurs at electric fields a little below the critical one where the mean electron energy increases sensitively with electric field. This sensitivity is sufficient to cause a negative-differential mobility for an electric field greater than 350 V mm^{-1} where the electrons are 'caught' by the satellite valleys in preference to energy runaway.

At high fields the velocity approaches an asymptote owing to drift-velocity saturation in the satellite valleys. Additional scattering mechanisms are available for 'hot' electrons, and they are responsible for the saturation effects (Bulman *et al.* 1972).

In summary, the material requirements for a significant transferred-electron negative-differential mobility are as follows.

1. $\mathcal{E}_s$, the satellite to central valley energy gap, must be several times larger than the thermal energy in order to avoid populating the satellite valleys at low electric field.

2. $\mathcal{E}_g$, the fundamental energy gap, must be greater than $\mathcal{E}_s$ in order to avoid impact ionization of electrons across $\mathcal{E}_g$ occurring before intervalley transfer.

3. The effective mass in the satellite valleys must be appreciably larger than that in the lower valley. Electrons with sufficient energy to have the choice of occupying either valley will then have a much greater probability of occupying the satellite valleys owing to their relatively high density of states.

4. The electron mobility in the satellite valleys must be much smaller than that in the lower valley.

5. Transfer of electrons between the valleys must occur over a small range of electric fields.

2.2 Internal instability and domain formation

The shape of the v–E characteristic suggests that a d.c. bias in the negative-differential mobility region would allow generation of a.c. power in a circuit connected to the device. This mode of operation is indeed possible and is called the LSA (*l*imited *s*pace-charge *a*ccumulation) mode. Stringent conditions are placed on its realization, and further description will be given later.

One difficulty in directly utilizing the device as an a.c. negative resistor arises from internal instabilities. A simplified example is shown in Fig. 2.4. The uniformly-doped device in Fig. 2.4 (a) has a spontaneous fluctuation of electron density which may be caused by a noise process or by a defect in the doping uniformity. There will be an electric-field non-uniformity related to the space charge by Poisson's equation

$$\epsilon\epsilon_0 \frac{\partial E}{\partial x} = e(n - N_D) \tag{2.5}$$

where N_D is the doping density. For convenience in eqn (2.5), the electron is considered to have a positive charge and the donors a negative one. This convention is commonly used in problems concerning electrons alone. If the mean electric field is below threshold, electrons in the perturbed region of higher field will move faster than the electrons elsewhere, and the space-charge accumulation tends to fill in the depleted region to restore neutrality. This is the normal ohmic process of space-charge damping by dielectric relaxation. If the mean electric field is above threshold there is a reduction of the electron-drift velocity in the region of higher field so that electrons in the partially depleted region fall back with respect to the travelling space-charge pattern and cause both an increase of depletion and accumulation at the appropriate points of the space-charge pattern. In turn, following eqn (2.5), there is an increase of the electric-field non-uniformity and a greater reduction of the electron velocity in the non-uniform region (Fig. 2.4(b)). An exponential growth of the space-charge non-uniformity occurs until the large-signal, non-linear, stable domain is formed, as shown in Fig. 2.4 (c). The domain becomes stable when the electric field E_R outside the domain has fallen below E_T and removed the drift-velocity difference which causes the the small-signal space-charge growth. A simple mathematical description of the above space-charge growth or decay processes is given in Chapter 3 from eqn (3.4) to eqn (3.8), where the characteristic time of growth or decay is identified as the dielectric relaxation time τ_ϵ.

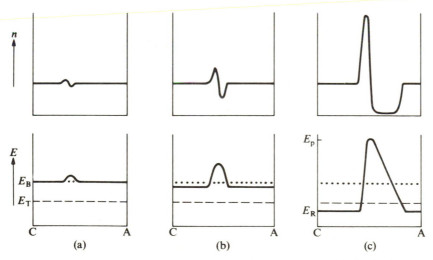

Fig. 2.4. Illustration of the growth of space-charge fluctuations to a stable dipolar domain. (a), (b), and (c) are sequential. C is the cathode and A is the anode.

2.3. Domain dynamics
The stable space-charge profile, or domain, drifts at constant velocity through

the device. On the cathode side, the electric field is below threshold. It rises steeply through the accumulation layer to a peak value in excess of E_T and decreases through the depletion layer to a value below threshold on the anode side. The stable form of this non-linear entity is dictated by the interplay of the growth and decay processes appropriate to the electric fields in different parts of the domain. Analysis of its form requires non-linear solution of the current-continuity equation and Poisson's equation subject to satisfaction of the $v-E$ characteristic and is a necessary key to understanding the current–voltage characteristic and other characteristics of a device containing a stable domain. The problem was first solved analytically in a most elegant manner by Butcher (1965) as outlined in Appendix I. The interested reader may find value in the further detailed analyses by Butcher, Fawcett, and Hilsum (1966), Butcher and Fawcett (1966), and Butcher, Fawcett, and Ogg (1967). For the immediate purposes two important results emerge. If the diffusion constant D of the electrons is assumed to be independent of the electric field the domain travels at the same velocity v_R as the electrons in the uniform-field region outside the domain. Furthermore, there is a characteristic curve which relates v_R to the peak field E_P and is called the dynamic characteristic (not to be confused with the later dynamic effects in microwave power generation). Appendix I shows that this curve is related to the $v-E$ curve by a simple geometrical construction; the two cross-hatched areas of Fig. 2.5 must be equal. Using these two results, it is possible to give a physical outline of the current conduction processes within the domain.

The current density J, must be continuous throughout the device, therefore

$$J = nev(E) + \epsilon\epsilon_0 \frac{\partial E}{\partial t} - eD \frac{\partial n}{\partial x}. \tag{2.6}$$

Outside the domain both n and E are independent of position so that the current is carried entirely as conduction current. The current density is nev_R, where v_R is the drift velocity corresponding to E_R. Within the domain all three processes

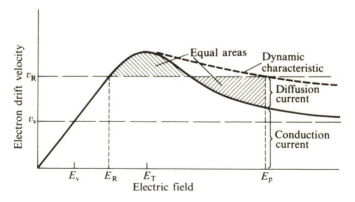

Fig. 2.5. Illustration of the equal-area relationship between E_p and E_R.

occur owing to gradients in n and E. The motion of the electric-field gradients causes displacement current to flow. Fig. 2.5 illustrates the identifiable electric fields in a domain. At the peak field E_p there is no displacement current because $\partial E/\partial x = 0$. However, there is an electron-density gradient which causes diffusion current to flow. The relative values of diffusion and conduction current at this point are indicated in Fig. 2.5. For the depletion layer the current is predominantly carried by displacement current while in the accumulation region there is a large conduction current opposed by displacement and diffusion currents. The electric-field and space-charge profiles are consistent with Poisson's equation. If the diffusion coefficient D is assumed to be zero, reference to eqn (AI.9) in Appendix I shows that the electron density within the domain ($v(E) \neq v_R$) must be zero (depletion region) or infinity (accumulation region). Outside the domain ($v(E) = v_R$) the electron density is equal to the doping density N_D. In this simple extreme the domain is triangular as shown in Fig. 2.6. The accumulation layer

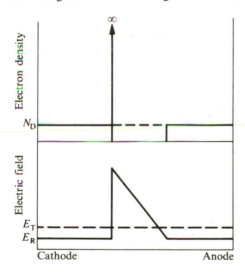

Fig. 2.6. Domain profiles in the limit of zero diffusion.

has zero width and infinite electron density so that the total accumulated charge is finite. The reader should note that the analysis cannot be carried out if the diffusion coefficient is initially assumed to be zero, owing to the presence of this charge singularity.

In the case of a fully depleted domain the equality of v_R and the domain velocity v_D is easily shown. In the completely depleted region

$$J = \epsilon\epsilon_0 \frac{\partial E}{\partial t} = -\epsilon\epsilon_0 \frac{\partial E}{\partial x} v_D.$$

The transferred-electron effect and space-charge instabilities

Poisson's equation requires

$$\epsilon\epsilon_0 \frac{\partial E}{\partial x} = -N_D e,$$

therefore

$$J = N_D e v_D. \tag{2.7}$$

The current outside the domain, where $\partial E/\partial x = 0$ and $n = N_D$, is

$$J = nev_R = N_D e v_R. \tag{2.8}$$

By current continuity, eqns (2.7) and (2.8) require

$$v_R = v_D.$$

In the case of a fully depleted domain the result is independent of the electric-field dependence of D.

The next question to answer is : what is the stable value of the outside field E_R? To aid us we need a convenient measure of the domain size. This is the domain potential ϕ_D, where

$$\phi_D = \int_{E > E_R} (E - E_R)\, dx. \tag{2.9}$$

A typical relationship between ϕ_D and E_R is illustrated in Fig. 2.7. The form of this relationship may be seen in the simple case where zero diffusion is assumed

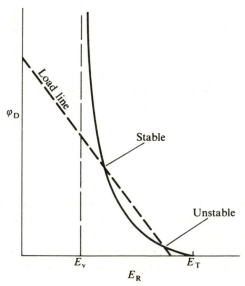

Fig. 2.7. The relationship between ϕ_D and E_R (solid line) imposed by the space-charge dynamics. Simultaneously the device must satisfy the boundary conditions contained in the load line.

so that the domain is fully depleted and triangular. Then

$$\phi_D = (E_p - E_R)\frac{w}{2},\tag{2.10}$$

where w is the domain width. From Poisson's equation

$$w = \frac{\epsilon\epsilon_0 (E_p - E_R)}{N_D e},\tag{2.11}$$

where $\epsilon\epsilon_0$ is the dielectric permittivity, therefore

$$\phi_D = \frac{\epsilon\epsilon_0 (E_p - E_R)^2}{2 N_D e}.\tag{2.12}$$

This relationship of ϕ_D and E_R (after using the dynamic characteristic to relate E_p and E_R) is dictated only by the space-charge dynamics and does not specify a single value of E_R. Using the above considerations the domain could be large and the outside field small or vice versa. In addition, the electrical boundary conditions must be satisfied. These require

$$V_B = \phi_D + E_R l,\tag{2.13}$$

where V_B is the terminal bias voltage. For constant V_B this relationship is plotted as a 'load line' on Fig. 2.7. The stable operating point of a domain is given by the upper intersection of the $\phi_D - E_R$ curve and this load line. The load line crosses the axes at $\phi_D = V_B$ and $E_R = V_B/l$. It can be seen that $E_R \simeq \frac{1}{2} E_T$ for a long device (steep load line) if the peak—valley ratio of the $v-E$ characteristic is 2. For such a device the domain velocity is half of the threshold drift velocity.

The $\phi_D - E_R$ curve also shows that a domain, once nucleated, can continue to propagate even if the terminal voltage is reduced below threshold. There is another characteristic voltage, called the sustaining voltage V_S, below which the domain will be extinguished in flight. At the sustaining voltage the load line is a tangent to the $\phi_D - E_R$ curve. For a long device this voltage is approximately half the threshold voltage and is important for later considerations of the cavity control of oscillators.

Some load lines have two intersections with the $\phi_D - E_R$ curve and the lower one in Fig. 2.7 is unstable. The reason can be seen by considering the effect of a noise fluctuation of ϕ_D in a domain operating at the unstable point. The load-line conditions must be satisified at all times, but the $\phi_D - E_R$ curve only refers to a steady-state domain. Accordingly, a small increase of ϕ_D causes E_R to decrease but become greater than dictated by the $\phi_D - E_R$ curve, and electrons on either side of the domain move faster than the domain. Both the total accumulation and depletion of electrons increase causing a further increase of ϕ_D (to satisfy Poisson's equation). This causes the working point to move further up from the unstable point and the regenerative readjustment continues until the stable working point is reached.

11

The transferred-electron effect and space-charge instabilities

Most of the gross features of domain behaviour are unaltered if the variation of D with E is included, except that the domain travels at a velocity which is a little different from the electron-drift velocity outside the domain (Butcher *et al.* 1967). The essential reason for this modification lies in the asymetry of diffusion currents caused by the asymetry of the electric field in a domain.

2.4. The current–voltage characteristic of a device containing a steady-state domain

In the previous section the load line was determined only by the device length and bias voltage, whereas the ϕ_D-E_R relationship of a steady-state domain was determined only by the donor-doping density, the $v-E$ relationship, and to a small extent by the $D-E$ relationship. Simultaneous satisfaction of both relationships gave a unique value of ϕ_D and E_R for the travelling domain. From these two quantities the device current density is

$$J = ne\mu E_R, \tag{2.14}$$

and the terminal voltage is

$$V_B = \phi_D + E_R l. \tag{2.15}$$

Eqns (2.14) and (2.15) allow the $J-V$ characteristic to be generated and some examples are given in Fig. 2.8. These characteristics are often used to describe

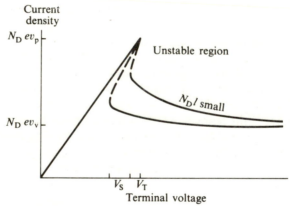

Fig. 2.8. The current density versus terminal voltage of devices containing a steady-state domain. The post-threshold characteristic with generally higher current has the smaller $N_D l$ product.

the behaviour of domain oscillators even when the terminal voltage is time-varying. It is assumed that the domain is in quasi-static equilibrium with the terminal voltage, but this condition will not be satisfied in many devices. In the case of the transit-time oscillators described in the next section the mean current is usually assumed to be equal to that given by eqn (2.14).

It is interesting to note that the large-signal negative conductance of transferred-electron devices is much the same for all the possible modes of operation (see Chapter 4 and 5). The reason for this may be seen by comparing the current—voltage characteristic of Fig. 2.8 with that implied by the v—E characteristic of Fig. 2.1 for conditions of electrical uniformity. Essentially, the mean large-signal current—voltage characteristic of any mode lies between these extremes of uniformity and maximum non-uniformity.

2.5. The domain transit-frequency

In a device with a finite length, a domain usually forms near to the cathode because this region often has the greatest non-uniformity owing to crystal damage occurring in the contacting process. The domains drift at a velocity of approximately 10^5 ms^{-1} (in a long device) until they enter the anode. As they do so, the 'absorbed' domain potential is redistributed into the field outside the domain. The current also rises with this field until it reaches threshold when it falls again as the next domain forms. The current through the device has a repetitive pulse form superimposed on a constant value as shown in Fig. 2.9. This

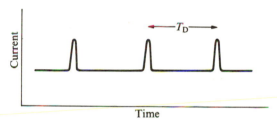

Fig. 2.9. A 'classical' transit-time current waveform of a Gunn device containing a steady-state domain.

is the form of the effect originally observed by Gunn (1963, 1964). For a device length of 100 μm and $N_D l$ product (see later) significantly greater than 10^{16} m^{-2}, the transit-frequency $1/T_D$ is ~ 1 GHz. Many practical devices have a length ~ 10 μm but their $N_D l$ product is close to 10^{16} m^{-2}. In this case the domain (or space-charge) velocity can be somewhat greater than 10^5 ms^{-1} and the transit-time is modified by the finite time of space-charge growth to a mature form. The transit-frequency will usually lie between 7 GHz and 25 GHz (Hobson 1967) for such a device , depending upon its bias voltage.

2.6. Experiments to verify the transferred-electron mechanism

We now turn briefly to the experimental evidence that the transferred-electron mechanism is indeed the one occurring in the Gunn effect and related devices. All materials that have exhibited the effect (gallium arsenide, indium phosphide, cadmium telluride, zinc selenide, and indium arsenide under pressure) have the same band structure as gallium arsenide. Other materials with the same conduction-band structure which do not exhibit the effect have a fundamental energy gap, between the conduction and valence bands, which is smaller than the energy gap between the central and satellite valleys. In these materials the approach to polar breakdown at high fields causes impact ionization of electrons across the fundamental energy gap instead of intervalley transfer (Foyt and McWhorter 1966). Indium arsenide is a material whose behaviour may be changed from the exhibition of impact ionization to the Gunn effect by application of uniaxial stress along the ⟨111⟩ direction (Allen, Shyam, and Pearson 1966). Such a stress modifies the band structure by lowering the satellite-valley energy with respect to the central valley and by increasing the fundamental energy gap. After the attainment of a critical pressure intervalley transfer occurs before impact ionization.

The crucial experiment which first demonstrated that the two-valley transfer model was correct was performed on gallium arsenide under hydrostatic pressure (Hutson, Jayaraman, Chynoweth, Coriell, and Feldman 1965). As the pressure was increased it was found that the threshold field for the Gunn effect decreased in accordance with the expected lowering of the satellite-valley energy with respect to that of the central valley. At a pressure of 26 kbar the Gunn-effect oscillations disappeared. This pressure is in good agreement with that theoretically required to cause the satellite-valley minima to fall to the same energy as the central-valley minimum. Under this condition the satellite valleys would be occupied at room temperature in the absence of an electric field. Therefore field-induced transfer could not take place and a negative-differential mobility would not exist.

2.7. Accurate calculation of the v–E characteristic

The predominant small-angle scattering of electrons in the central valley of the conduction band of GaAs by polar lattice vibrations causes their velocity (or wave-vector) distribution to be highly non-Maxwellian. Under hot-electron conditions the distribution function becomes elongated in the drift direction and has provided many theoretical difficulties in the claculation of the v–E characteristic (Bulman et al. 1972). The displaced Maxwellian (Butcher 1967; Bott and Fawcett 1968) and relaxation time (Conwell and Vassell 1968) approximations were used initially, but they have weaker justification than they did for hot-electron descriptions of germanium or silicon (Paige 1964; Conwell 1967). Modern computers have enabled the successful development of Monte Carlo numerical techniques (Fawcett, Boardman, and Swain 1970) to overcome these difficulties. The movement of a single electron through velocity (or wave-vector) space is logged while random numbers with the appropriate probability distributions determine the interval between, and the change of velocity caused by, collisions. Any average over the velocity distribution may be simply summed without analytical approximation. Ruch and Fawcett (1970) used these techniques to great practical value in calculating the dependence of the v–E relationship on temperature and impurity concentration as shown in Figs. 2.10 and 2.11 respectively.

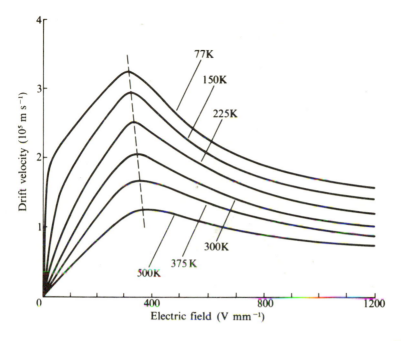

Fig. 2.10. Temperature-dependence of the velocity—field characteristic in pure GaAs (Ruch and Fawcett 1970).

At low temperature and low electric field the electron mobility is notably high in pure material. Under these conditions the mean electron energy is appreciably smaller than the quantized energy which must be emitted as a polar-optical phonon when an electron loses energy on collision with the crystal lattice. Only those electrons with high energy in the 'tail' of their energy distribution can take part in the energy-loss process.

2.8. Measurement of the $v-E$ characteristic

The instabilities described in § 2.2 present considerable difficulty for measurement of the $v-E$ relationship. As will be seen in Chapter 3 these instabilities only occur if the product of donor-doping density and device length exceeds a critical value. For smaller products the device is stable but will often have a highly non-uniform electric-field distribution. Measurement techniques must take account of these difficulties, which preclude determination of the shape of the $v-E$ curve by a simple measurement of the terminal current—voltage characteristic. Measurements are usually made in a time which is short compared with the formation time of the non-uniformities (§ 3.2) or they are made on sub-critically doped devices (Bulman *et al.* 1972).

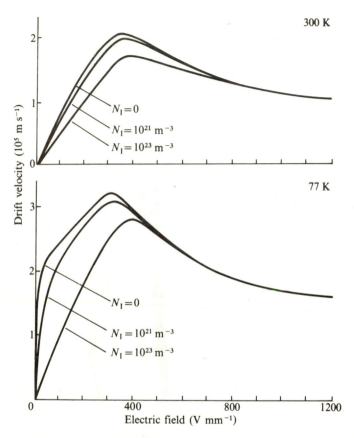

Fig. 2.11. The effect of ionized-impurity scattering by donors whose concentration is the curve parameter N_I (Ruch and Fawcett 1970).

In the former case the measurement of the current–voltage characteristic of a device may be made with fast-pulse techniques (Gunn and Elliot 1966). An alternative way of generating high-speed excursions into the negative-differential mobility region is through the use of high-power microwave signals applied to the device across the centre of a waveguide. This so-called 'microwave-heating' technique requires measurement of the relationship between the mean current and voltage through and across the device in the presence of a large microwave signal whose electric-field amplitude is greater than the threshold field. The d.c. components are generated by the non-linearity of the v–E characteristic, which is derived by finding the curve which gives the best fit to the experimental data (Glover 1971). Care must be taken when choosing the microwave frequency both to avoid formation of non-uniformities, at one extreme, and to allow the electron momentum distribution sufficient time to reach equilibrium, at the other

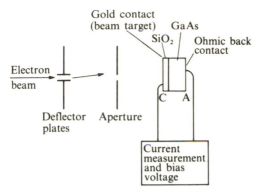

Fig. 2.12. Schematic outline of the equipment used by Ruch and Kino (1967, 1968).

extreme (Bulman *et al.* 1972).

Measurements on subcritically doped devices fall into two categories. The non-uniform electric-field distribution in amplifier diodes described in Chapter 5 may be measured with a point-contact probe and related to the $v-E$ curve (Thim 1966). Problems exist in obtaining sufficient spatial resolution. The second category of static-field measurements involves measurement of the space-charge transit-time in a depleted diode with a barrier contact at the cathode. Ruch and Kino (1967, 1968) carried out the first successful measurements on semi-insulating GaAs using this technique, which gives the most direct measurement of velocity and electric field. Their diodes had a cathode barrier consisting of a thin layer of evaporated silicon dioxide on the intrinsic GaAs, with a thin gold contact evaporated onto the silicon dioxide. Both layers had a thickness less than 100 nm. A very short pulse (~ 0.1 ns) of electrons was directed on to the gold contact by rapidly sweeping an electron beam past a small aperture in line with the device (Fig. 2.12). The injected pulse of electrons multiplied on impact (approximately 1000 times) and then drifted under the influence of a reverse electric field towards the anode contact. Current flow was observed in the external circuit for the duration of the electron flight from cathode to anode, so allowing it's timing and velocity measurement (Fig. 2.13). Good agreement was found between the measured and theoretically predicted characteristics (Ruch and Fawcett 1970). A later development of this technique (Evans, Robson, and Stubbs 1972) using microwave phase as a measure of the time of flight looks promising for measurements on the short devices, with lower resistivities, that are used for practical oscillators and amplifiers. None of the techniques have given results at high fields ($> 10\,000$ V mm^{-1}), and inaccurate knowledge of the appropriate electron-transport parameters has not allowed reliable predictions to be made of this asymptotic part of the characteristic. Experiments with GaAs IMPATT devices suggest that a velocity of 0.65×10^5 m s^{-1} at 25 000 V mm^{-1} is appropriate at room temperature.

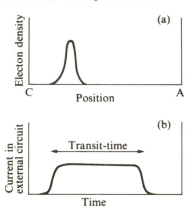

Fig. 2.13. (a) Injected electron bunch in the experiment of Ruch and Kino (1967, 1968). (b) The resulting current in the external circuit allowing transit-time measurement.

2.9. Intervalley scattering and energy relaxation

In the earlier discussion of the transferred-electron effect consideration was not given to the finite time of transfer from the central valley to the satellite valleys. This time will set an upper frequency limit to the transferred-electron effect because a velocity decrease with increase of field cannot occur until intervalley transfer has occurred. To illustrate these effects it is convenient to consider the response of electrons to a sinusoidal electric field which is assumed uniform and has a mean value in excess of the threshold field while the minimum value lies below threshold. When the electric field rises above threshold and the electrons in the central valley have sufficient energy to transfer to the satellite valleys they require a collision with the appropriate lattice vibration (Bulman *et al.* 1972) in order to provide the change of electron momentum dictated by the band structure of Fig. 2.2. The mean time between collisions for this process is approximately 10^{-12} s for a lattice temperature of 300 K, so that an upper frequency-limit may be expected at approximately 150 GHz, owing to intervalley scattering.

A more serious limitation arises in the time required for a change of the mean electron energy in the central valley. The relatively long time required for energy relaxation or electron 'thermalization' in the central valley has already been remarked upon in § 2.1. When the electric field is rising this process is not so important because all the electrons are accelerated by the electric field and can rapidly gain sufficient energy to allow intervalley transfer. Problems arise when the electric field is falling and electrons are required to transfer back to the central valley from the satellite valleys. Owing to the much greater density of states in the satellite valleys, a significant number cannot do this until the 'average' electron has lost sufficient energy in the central valley to preclude occupation of the satellite valleys. The time required is the energy-relaxation time and is approximately 3×10^{-12} s, so that serious limiting effects become apparent between 50 GHz and 100 GHz. The time-averaged negative conductance becomes weaker under these conditions because the terminal current of the device cannot rise as the electric field falls. A typical trajectory of electron-drift velocity and electric field is illustrated in Fig. 2.14 for the assumed uniform-field

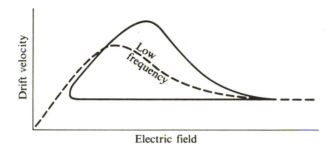

Fig. 2.14. Illustration of phase lag between the electron-drift velocity and electric field when the angular frequency is comparable with the inverse of the electron energy relaxation time in the central valley.

conditions when the frequency causes these limitations. Quantitative calculation of these effects has been carried out by Rees (1969) with an alternative computer method to the Monte Carlo technique described in § 2.7. It avoids solution of the Boltzmann equation (Bulman *et al.* 1972) by exploiting the stability of the

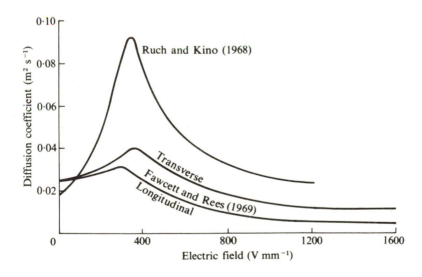

Fig. 2.15. The electron diffusion coefficient in GaAs as calculated by Fawcett and Rees (1969) both along and transverse to the electric field. An indirect experimental evaluation of the longitudinal-diffusion coefficient by Ruch and Kino (1968) is shown for comparison.

19

steady-state distribution function of the electron momentum. Momentum (or wave-vector) space was divided into finite cells and, at the start of the calculation, an arbitrary distribution of electrons was introduced over these cells. From each cell in turn, the electrons were scattered to all other cells according to a chosen time-step and the appropriate probabilities (Bulman *et al.* 1972). Under static-field conditions the process may be iterated until the form of the distribution function becomes invariant. Owing to it's calculation of the collision events and velocity redistribution of all electrons in one time increment, the technique naturally lends itself to time-varying problems where the distribution function depends on its history and is not in the steady state.

2.10. Diffusion

Electrons in either the central valley or the satellite valleys diffuse in the normal fashion, but we may expect the light electrons in the former valley to have a higher diffusion coefficient than the heavy electrons of the satellite valleys, owing to their higher mobility. When the electric field is such that there are a significant number of electrons of each type an additional diffusion-like process occurs. The effects may be illustrated by reference to Fig. 2.4 (c) as they would occur in the accumulation region of a domain. A light electron which is at the cathode side of the accumulation layer, where the electric field is below threshold but a little greater than the field outside the domain, will drift faster than the domain. It will eventually exceed the threshold field as it travels through the domain, and the probability of intervalley transfer will become progressively greater. When the electron becomes heavy, following transfer, it will drift more slowly than the domain and will be overtaken by the lower-field regions. When it lies in a field below threshold the electron will transfer back to its light form in the central valley after the appropriate time interval for intervalley scattering and energy relaxation. If the over-all scattering time in both directions between the two states was zero the domain shape would be dictated by the domain dynamics derived from the velocity—field characteristic. When the scattering time is appreciable the accumulation layer is broadened and the domain shape is modified (Szekely and Tarnay 1968; Cheung and Hearn, 1972). Effectively, an extra diffusion effect is present and is indicated by the peak diffusion occurring near the threshold field in Fig. 2.15.

3.1. Introduction

In the previous chapter the description of electron transfer and domain formation paid scant attention to the effect of boundary conditions in space and time. For example, domain growth to maturity takes a finite time which will depend on the size of its nucleating centre. If the device is short enough maturity will not be achieved before the space charge enters the anode. Effects such as these are described in this chapter in order to set a background for their various effects in oscillators and amplifiers.

3.2. The doping–length ($N_D l$) product

It may be expected that there will be a minimum device length, for a given doping density, that will support a domain, owing to the finite time required for domain growth. If this time is longer than the space-charge transit-time through the device, domain formation cannot occur. A simple, though imprecise, way of estimating this length is to equate it to the domain length under a threshold-bias voltage. Shorter devices simply cannot contain a domain. Following the guidance of Chapter 2 for a fully depleted domain, eqns (2.10) and (2.11) give

$$\phi_D = \frac{N_D e w^2}{2\epsilon\epsilon_0}, \tag{3.1}$$

where N_D is the doping density and w is the domain width. If the peak–valley ratio is 2 and we assume that the field outside the domain is half the threshold field we have

$$\phi_D = \tfrac{1}{2} V_T = \tfrac{1}{2} E_T w . \tag{3.2}$$

from eqns (3.1) and (3.2) the device may contain a domain providing that

$$N_D l > \frac{\epsilon\epsilon_0 E_T}{e} = 2 \cdot 4 \times 10^{14} \, \text{m}^{-2} . \tag{3.3}$$

The above estimate takes no account of the time taken for the space charge to grow to maturity. Furthermore, space-charge growth may be impossible. In order for instability to occur under, for example, constant -voltage conditions it is necessary for the small-signal equivalent circuit to contain a reactance which has a short-circuit resonance at a frequency where the series resistance is negative. These stability considerations were first made by McCumber and Chynoweth (1966), whose technique may be illustrated by the following small-signal solution of current-continuity and Poisson's equations. We will neglect diffusion and assume uniformity of the static electric field and space charge. The relevant equations are

$$J = nev(E) + \epsilon\epsilon_0 \frac{\partial E}{\partial t}, \tag{3.4}$$

where J is the current density. Poisson's equation requires

$$\epsilon\epsilon_0 \frac{\partial E}{\partial x} = (n - N_D)e. \tag{3.5}$$

In the small-signal stage of the growth eqns (3.4) and (3.5) may be linearized by a first-order Taylor expansion that separates the static quantities, denoted by subscript 0, and the dynamic small-signal quantities denoted by a subscript 1

$$E = E_0 + E_1,$$

$$n = N_D + n_1,$$

$$J = J_0 + J_1,$$

$$v = v_0 + \left(\frac{\partial v}{\partial E}\right)E_1.$$

Eqns (3.4) and (3.5) must be separately true for the static and small quantities,

$$J_1 = n_1 e v_0 + N_D e \left(\frac{\partial v}{\partial E}\right)E_1 + \epsilon\epsilon_0 \frac{\partial E_1}{\partial t},$$

$$\epsilon\epsilon_0 \frac{\partial E_1}{\partial x} = n_1 e.$$

Eliminating n_1 from these equations,

$$J_1 = \epsilon\epsilon_0 \left(v_0 \frac{\partial E_1}{\partial x} + \frac{\partial E_1}{\partial t}\right) + N_D e \left(\frac{\partial v}{\partial E}\right)E_1. \tag{3.6}$$

To maintain current continuity, J_1 is only a function of time. It can be verified by substitution that the solution of eqn (3.6) is

$$E_1 = \exp(-t/\tau_\epsilon).f(x - v_0 t) - f(t). \tag{3.7}$$

The last term in eqn (3.7) is simply a 'background' uniform-field term which follows the differential conductance of the $v-E$ characteristic. Its magnitude relative to the first term is dictated by boundary-condition requirements on E_1.

The effect of spatial and temporal boundaries

The first term on the right-hand side of eqn (3.7) shows that the electric-field (and space-charge) profile propagates at velocity v_0. It has a decay time constant, the dielectric relaxation time, which is given by

$$\tau_\epsilon = \epsilon\epsilon_0/N_D e(\partial v/\partial E) . \tag{3.8}$$

If $\partial v/\partial E$ is positive τ_ϵ is also positive and the space charge decays. However, a negative-differential mobility causes τ_ϵ to be negative and growth occurs. In order to determine the small-signal equivalent circuit the boundary conditions must be used to determine the relative magnitude of $f(t)$ and $f(x - v_0 t)$ in eqn (3.7). For example, ohmic contacts require that $E_1 = 0$ at both the anode and cathode. In turn, J_1 may be evaluated by eliminating E_1 between eqns (3.6) and (3.7). The form of the small-signal impedance is then given by

$$Y = \int_0^l E_1 \, dx/J_1 A , \tag{3.9}$$

where A is the device cross-sectional area and l is its length. Precise details of the frequency-dependence have been omitted from eqn (3.9), and the calculation is usually tedious and involved owing to spatial dependence of the static quantities (Holstrom 1967). McCumber and Chynoweth (1966) assumed uniform field conditions to show that instability will occur if

$$N_D l > 2 \cdot 09 \epsilon\epsilon_0 \Big/ e \left| \frac{1}{v_0} \left(\frac{\partial v}{\partial E} \right) \right| . \tag{3.10}$$

Insertion of the parameters for maximum negative $\partial v/\partial E$ in eqn (3.10) using the Ruch and Fawcett (1970) v–E characteristic at 300 K shows that the *critical $N_D l$ product is* $\sim 5 \times 10^{15}$ m^{-2}. An alternative interpretation of eqn (3.10) is that instability will occur if the sample length is greater than $2 \cdot 09$ negative dielectric relaxation lengths ($v_0 \tau_\epsilon$ is the dielectric relaxation length). Diodes with a shorter length are stable and may be used for amplification (see Chapter 5).

Even though eqn (3.10) gives the criterion for instability it is not a criterion for domain formation, which may take many dielectric relaxation times. This problem requires knowledge of the nucleating-region dimensions (Hobson 1969a) and the growth rate under large-signal conditions (see later), but if domain growth is completed within the device length we have the knowledge that the domain will 'fit' into the device (eqn (3.3)) when the bias voltage is close to threshold.

3.3. Effect of diffusion – the $N_D l^2$ product

The 'bunching' of space-charge accumulation or depletion in a medium with a negative-differential mobility is opposed by the spreading effect of diffusion. If the spatial scale of the fluctuation is too small ($\sim$ Debye length) diffusion damping will be large enough to inhibit space-charge growth. Small-signal calculations (Bulman *et al.* 1972; Ridley 1966), following the technique of the previous section but including diffusion, show that instability will only occur if

$$N_D l^2 > \epsilon \epsilon_0 D \pi^2 \left/ e \left(\frac{\partial v}{\partial E} \right) \right. , \tag{3.11}$$

where D is the diffusion coefficient. If $|\partial v / \partial E| = 0 \cdot 2$ m^2 v^{-1} s^{-1} and $D = 0 \cdot 02$ m^2 s^{-1}, inequality (3.11) becomes

$$N_D l^2 > 0 \cdot 69 \times 10^9 \text{ m}^{-1} .$$

The $N_D l^2$ instability requirement is always satisfied for devices with a length greater than $\sim 0 \cdot 15$ μm if the $N_D l$ requirement is satisfied. This short distance compared with usual device dimensions indicates that diffusion only 'rounds the corners' of calculations carried out by neglecting it.

3.4. Lateral effects – the $N_D d$ product

In all the previous considerations there has been an implicit assumption of uniformity perpendicular to the direction of space-charge propagation. The lateral dimensions of the space charge also influence its growth rate (Kino and Robson 1968). If they are not much greater than the longitudinal dimensions the electric-field lines of force tend to spread out radially from the charge (Fig. 3.1). The

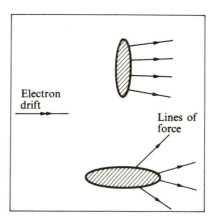

Fig. 3.1. The effect of transverse dimensions on the longitudinal electric-field difference across the same quantity of space charge in two different orientations.

electric field in front of the space charge of Fig. 3.1 is greater than that to the rear, but is more so in the upper configuration. The radial divergence present in the lower diagram reduces the electric-field difference across the space charge

The effect of spatial and temporal boundaries

when compared with a one-dimensional situation. There is a corresponding reduction of the electron-drift velocity difference which causes space-charge bunching and growth. Small-signal calculations, which become more tedious in two dimensions, show that instability will not occur in an infinitely long device with semi-insulating GaAs on either side of the active region if (Kino and Robson 1968)

$$N_D d < 1 \cdot 6 \times 10^{15} \, \text{m}^{-2}, \tag{3.12}$$

where d is the device width.

These effects are enhanced by the presence of high-permittivity dielectrics or metal conductors (Kataoka, Tateno, and Kawashima 1969) on the sides of the device (Fig. 3.2).

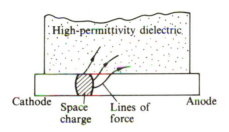

Fig. 3.2. Enhancement of the divergence of electrical lines of force by surface loading with a high-permittivity dielectric. A similar effect occurs when metals are brought close to the sides of a device.

3.5. Space-charge control by subcritical time

In the earlier consideration of the transferred-electron effect, the negative-differential mobility of the $v–E$ characteristic looked attractive as the basis of a frequency-independent negative conductance for use in amplifiers or oscillators. This hope was thwarted by the internal instability caused by a negative-differential mobility. The length-stabilization criteria discussed in § 3.2 suggest that the ideal negative resistor may be realized with subcritically doped devices ($N_D l \lesssim 5 \times 10^{15} \, \text{m}^{-2}$), but it will be seen in Chapter 5 that the negative-differential mobility causes 'space-charge masking' and the terminal conductance is positive except at frequencies which are again closely related to the space-charge transit-frequency. The next attempt to realize the bulk negative resistor only allows the electric field to exceed threshold for a time shorter than necessary to allow significant space-charge formation. This mode of operation is only possible when the a.c. terminal voltage is large enough to fall below threshold once per cycle; this is the LSA mode of operation (see § 2.2). The large-signal operation can only be achieved in a tuned circuit and is qualitatively illustrated in Fig. 3.3. As the mean electric field rises above threshold, space-charge inhomogeneities

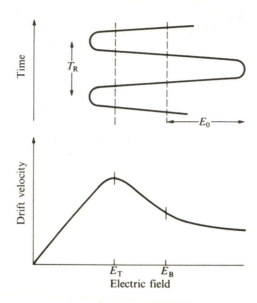

Fig. 3.3. LSA operating conditions.

begin to form. Before they have grown to a significant size the electric field must become large enough for the instantaneous negative-differential mobility to fall to a very low value, so that the dielectric relaxation time is very long at the turning point of the sinusoidal electric field. The radio frequency (r.f.) period T_R must be less than the dielectric relaxation time averaged over the time that the electric field exceeds threshold if the space-charge perturbation is to be limited to a small value. When the electric field falls below threshold it is necessary for the incipient space charge to be damped by the positive mobility. This condition may be conveniently written as $T_R \gg \tau_{\epsilon_0}$, where τ_{ϵ_0} is the low-field dielectric relaxation time. In order to avoid successive build-up of small amounts of space charge over many cycles it is necessary for the positive-differential mobility damping to exceed the negative-differential mobility growth. In addition, any injected space charge (such as the accumulation layers or inhomogeneity effects described later) must be damped. More detailed consideration (Copeland 1967c; Bulman *et al.* 1972) show that the growth and decay conditions require

$$\frac{\epsilon\epsilon_0}{ne|\bar{\mu}_n|} > T_R \gg \frac{\epsilon\epsilon_0}{ne\mu_0},$$

which approximately reduces to

$$2 \times 10^{11} > n/f > 2 \times 10^{10} \text{s m}^{-3}, \tag{3.13}$$

with $\epsilon = 12\cdot5$ and $|\bar{\mu}_n| = 0\cdot03 \text{ m}^2 \text{ v}^{-1}\text{s}^{-1}$.

The effect of spatial and temporal boundaries

Further discussion of LSA devices as oscillators will be postponed until the next chapter.

3.6. Space-charge injection from doping non-uniformities

In the previous sections of this chapter consideration has been given to the effect of spatial and temporal boundary conditions on the growth of space charge. Attention is now turned to the initial conditions which nucleate the space charge. In the bulk of ideally uniform material, statistical fluctuations in the electron density or doping density will set the starting point (Hobson 1969a). In real materials we may expect larger fluctuations in the doping uniformity owing to technological limitations. Fig. 3.4 illustrates the nucleating effect of the static-doping non-uniformities. The random motion of electrons is neglected, and they are considered as a continuous 'sea' of charge. In the absence of an electric field, the electron-density profile would be the same as the doping profile, to ensure charge neutrality. On application of an electric field the electrons drift from cathode to anode and shift their profile as shown by the dashed line. In a positive-resistance medium the resulting space charge would be readjusted towards neutrality. In a medium with a negative-differential mobility the space charge grows so that the doping inhomogeneity forms the effective space-charge profile from which growth occurs.

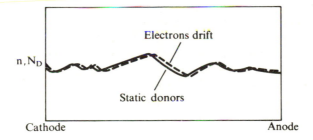

Fig. 3.4. Nucleation of space charge by a doping inhomogeneity.

Gross doping non-uniformities occur by necessity at the anode and cathode. The former does not contribute to dynamic instabilities because electrons are running into it. If the cathode is an ohmic contact it causes accumulation-layer injection each time that the electric field in the cathode region rises above threshold. A simple example illustrates accumulation-layer formation in Fig. 3.5. A step function of electric field is applied to the device causing it to exceed threshold. The doping (solid line) is uniform except for a highly doped, step contact. Just after application of the electric field the electron density and field distribution are illustrated by the dotted line. The maximum electron-drift velocity will occur where the electric field is equal to the threshold field E_T. This non-uniformity of

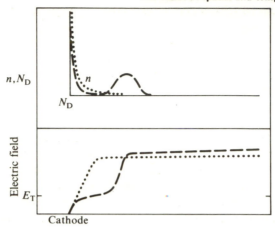

Fig. 3.5. Illustration of accumulation-layer formation.

velocity causes a bunching of electrons, as shown by the dashed line in Fig. 3.5, and a corresponding, space-charge induced, field reduction behind them. The process carries on until the electric field in front of the accumulation layer is above threshold and causes an electron velocity which is equal to that behind the layer where the electric field has the appropriate value below threshold.

Unlike the dipolar domain, the accumulation layer is not stable in an inhomogeneous medium. Space charge clearly cannot grow on the cathode side of the accumulation layer, but it will do so on the anode side where the electric field lies in the region of negative-differential mobility. If it is given sufficient time, the accumulation layer is converted to a dipolar domain (Kroemer 1966; McCumber and Chynoweth 1966) because the upward curvature of the v–E characteristic at high fields causes field-reducing fluctuations to grow more rapidly than those which increase it.

3.7. Domain and accumulation-layer relaxation rates

Having identified the small-signal growth rates and injection mechanisms of space charge it is only necessary to identify the readjustment rates of domains and accumulation layers before we can consider their combined effects in oscillators. In the large-signal limit, eqns (3.4) and (3.5) are difficult to solve owing to the non-linearity of the v–E relationship. The simplest course of action for a domain is to assume that the depletion region of the domain acts as a capacitor C_D while the region outside the domain is a series resistor R_0 (Hobson 1966a). The domain relaxes to equilibrium at a rate determined by the time constant $R_0 C_D$. Some justification for this approach may be obtained from more detailed considerations of domain relaxation (Kurokawa 1967). The domain capacity

The effect of spatial and temporal boundaries

may be calculated from the charge on the domain and the domain potential (Hobson 1966b). For a fully depleted, triangular domain the capacity is given by

$$C_D = \epsilon\epsilon_0 A/w, \tag{3.14}$$

where w is the domain width and A is the area. The series resistor is

$$R_0 = l/N_D e\mu A . \tag{3.15}$$

l is the device length, and it is assumed that $l \gg w$.

The domain relaxation time is

$$\tau_s = R_0 C_D = \frac{l}{N_D e\mu} \cdot \frac{\epsilon\epsilon_0}{w} . \tag{3.16}$$

From eqn (3.1), the domain potential ϕ_D is related to the domain width of a fully depleted, triangular domain by

$$\phi_D = \frac{N_D e w^2}{2\epsilon\epsilon_0} ,$$

therefore

$$\begin{aligned}
\tau_s &= \left(\frac{\epsilon\epsilon_0}{N_D e\mu} \cdot \frac{l^2}{2\mu\phi_D} \right)^{\frac{1}{2}} \\
&= \left(\frac{\epsilon\epsilon_0}{N_D e\mu} \cdot \frac{l}{\mu E_R} \cdot \frac{E_R l}{2\phi_D} \right)^{\frac{1}{2}} , \\
&= \left(\tau_{\epsilon_0} \cdot T_D \cdot \frac{E_R l}{2\phi_D} \right)^{\frac{1}{2}} , \tag{3.17}
\end{aligned}$$

where τ_{ϵ_0} is the low-field dielectric relaxation time and T_D is the domain transit-time. For a long device biased just above threshold $E_R \simeq \frac{1}{2} E_T$, therefore

$$(\tau_s)_{V=V_T} = (\tfrac{1}{2} \tau_{\epsilon_0} T_D)^{\frac{1}{2}} . \tag{3.18}$$

Material with an electron density of $10^{21}\,\mathrm{m^{-3}}$ has $\tau_{\epsilon_0} \simeq 10^{-12}\,\mathrm{s}$. A device with a length of $100\,\mu\mathrm{m}$ has $T_D \simeq 10^{-9}\,\mathrm{s}$, therefore

$$(\tau_s)_{V=V_T} \simeq 2\cdot25 \times 10^{-11}\,\mathrm{s}.$$

The decay or growth of an accumulation layer is much more rapid than that of a domain, which makes it much less lethal to LSA operation (Bulman *et al.* 1972; Hobson 1972b). In essence, the rapid readjustment arises from the unsymmetrical distribution of electric field on either side of the accumulated charge. In non-equilibrium conditions there may be a large disparity between the electron velocities on either side of the charge layer, so that the rate of flow of charge to or from it can be very high. This possibility does not exist for a domain.

4 | Cavity-controlled oscillators

4.1. Introduction

When a transferred-electron device is placed in a cavity or resonant circuit several modes of operation are possible. As shown in Chapter 7 the cavity or circuit may be represented in the form of Fig. 4.1 for single-frequency operation. The inductance L and capacity C_c comprise the resonant circuit, and G_L is the circuit load representing spurious losses and useful r.f. power dissipation. The various modes of operation depend upon (1) the relative values of T_R, the resonant period of the circuit, and the space-charge or domain transit-time T_D;

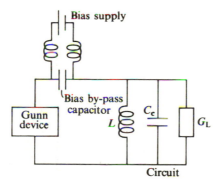

Fig. 4.1. The equivalent circuit of a Gunn oscillator in a circuit with a single resonant frequency. The bias supply is short-circuited to r.f. signals by the by-pass capacitor.

(2) the relative values of T_R and the relaxation time of space charge within the device; (3) the circuit loading G_L; and (4) the doping uniformity of the device. For the moment we will make some simple but drastic assumptions. The capacity directly across the device terminals or within the device itself will be assumed to have an impedance which is much smaller than the parallel negative resistance, so that the circuit will cause a sinusoidal voltage with amplitude V_0 to be impressed upon the bias voltage V_B at the device terminals. When this capacity restriction is removed non-sinusoidal waveforms are possible and operation of the form described in § 4.10 may take place. The relaxation times for space-charge growth or decay will be assumed initially to be instantaneous. Several possible modes of operation under these conditions are illustrated in Fig. 4.2,

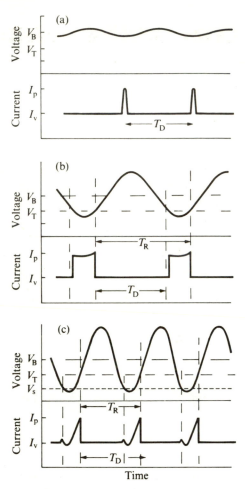

Fig. 4.2. Illustration of the terminal conditions for transit-mode (a), delay-mode (b), and quench-mode (c) operation of a Gunn oscillator.

where V_B is greater than the threshold voltage V_T. Particular advantages or disadvantages of any mode from the point of view of tunability and stability will be discussed at the appropriate point of later chapters.

4.2. The transit-time mode
In a purely resistive circuit ($L \to \infty$, $C_c \to 0$) the frequency is determined only by the space-charge transit-time across the device, and we have transit-mode operation. As shown in Fig. 4.2 (a) the current spikes occur when a domain enters the anode and the next one is nucleated from the cathode (see also § 2.5). If the

circuit is heavily loaded (i.e. G_L is large) the r.f. voltage amplitude is small and will not swing below V_T. The precise phase of the voltage waveform with respect to the current waveform will be determined by the values of L, C_c and G_L and the transit-frequency. Fourier analysis of the current pulse would give a component, at the fundamental frequency, in antiphase with the r.f. voltage, because the current is high when the voltage is low, and vice versa. This indicates that the device is delivering power to the load and it has a negative conductance. The small a.c. voltage amplitude and the narrowness of the current pulse suggest that the efficiency of this mode will be low. A further disadvantage is the limitation of the frequency to the natural domain transit-frequency.

4.3. The delayed-domain mode

For this mode the circuit Q-factor and $1/G_L$ must be large enough to allow a sinusoidal voltage waveform whose amplitude is large enough, at the device terminals, to cause the voltage to fall below threshold over a portion of each cycle. The domain transit-time must be less than T_R, so that the domain may disappear into the anode while the voltage is below threshold. The next domain is not nucleated until the voltage once more rises above threshold. During this delay time the diode behaves as a resistor with its low-field value. The waveforms are shown in Fig. 4.2 (b). Once again, it can be seen that Fourier analysis would reveal a component of current in antiphase with the voltage. The efficiency is higher than for the transit-time mode since the mark—space ratio of the current waveform is much larger. Up to $7 \cdot 2$ percent efficiency may be obtained if $I_p/I_v = 2$ (Warner 1966). A further advantage is the frequency controlability by the circuit, subject to the approximate limits

$$2T_D > T_R > T_D. \tag{4.1}$$

4.4. The quenched-domain mode

When the circuit loading is further reduced (G_L smaller) it may become possible for the terminal voltage to fall below the domain-sustaining voltage for a portion of each cycle. As the voltage falls below V_s the domain is quenched in flight and the next one is not nucleated until the terminal voltage again rises through threshold. Frequencies higher than the transit-time frequency can be generated. The high-frequency limit for this mode is not infinity, as may be inferred from the above description; it is limited by the time τ_s taken for domain nucleation or extinction (see § 3.7). The frequency constraints are approximately given by

$$2T_D > T_R > \tau_s \tag{4.2}$$

so that frequency control by the circuit is possible over a greater range than that for the delay mode. Within the limits of the delay mode given by inequality (4.1)

33

it may be possible to have either quench or delay modes, depending on G_L and the voltage amplitude. Quench-mode operation will occur if the voltage falls below V_s before the domain enters the anode, or vice versa. Comparison of the area of the current waveforms lying between I_p and I_v in Figs. 4.2(b) and 4.2(c) indicates that the quench mode will not be as efficient as the delay mode and calculations usually predict a maximum efficiency of about 5 percent.

4.5. The Pf^2 product

In combination with restrictions imposed by practical circuits the power output of any transit-time oscillator suffers a fundamental limitation for the reasons enumerated below.

1. Operation of domain-mode transit-time oscillators is limited to approximately on octave frequency range centred on the transit-time frequency, so that the frequency f is related to the length l by

$$l \simeq v/f, \tag{4.3}$$

where v is the domain velocity ($\sim 10^5 \, \text{m s}^{-1}$).

2. The maximum permissible electric field within the device is the avalanche-breakdown field E_B. The peak—peak r.f. electric field must not exceed E_B even in the most optimistic conditions of domain oscillation. Therefore, the maximum possible r.f. voltage amplitude is given approximately by

$$V_{0m} \simeq \tfrac{1}{2} E_B l . \tag{4.4}$$

3. In order to satisfactorily couple the microwave power out of a device and into the useful circuit the device impedance Z must not fall below a certain level which is not precisely defined but is of the order of $1 \, \Omega$ to $10 \, \Omega$ (see § 7.3). As the sample length is decreased for high-frequency operation, the area must also be decreased to satisfy this impedance limitation, but the precise details depend on the material-doping requirements. If i_0 is the r.f. current amplitude

$$V_{0m}/i_0 \simeq |Z| . \tag{4.5}$$

If i_a is the r.f. current component in antiphase with the r.f. voltage the microwave power output is

$$P = \tfrac{1}{2} V_{0m} i_a . \tag{4.6}$$

4. In order to satisfy thermal design requirements for continuous wave (c.w.) operation or to obtain a good depth of current modulation in pulsed operation it is necessary to work with a doping × length product which is substantially frequency-independent (see § 6.3). In turn, this implies that the ratio of conductive to susceptive current through well-designed devices is substantially constant over a wide range of frequencies (see § 7.3) and

$$i_\mathrm{a} = \frac{i_0}{\sqrt{(1 + Q_\mathrm{D}{}^2)}}, \tag{4.7}$$

where Q_D is the diode Q-factor giving the ratio of its susceptive to conductive currents.

Combining eqns (4.3), (4.4), (4.5), (4.6), and (4.7) we have

$$Pf^2 = \frac{1}{|Z|} \frac{E_\mathrm{B}{}^2 v^2}{8\sqrt{(1 + Q_\mathrm{D}{}^2)}} \tag{4.8}$$

or

$$P|Z|f^2 = \frac{E_\mathrm{B}{}^2 v^2}{8\sqrt{(1 + Q_\mathrm{D}{}^2)}}. \tag{4.9}$$

The Pf^2 relationship in eqn (4.8) shows that transit-time devices have a fundamental limitation of their power output. The $P|Z|f^2$ form in eqn (4.9) is often preferred because all the quantities on the right-hand side are properties of the device alone so allowing direct comparison with other types of transit-time device. Typical values of the material parameters are $E_\mathrm{B} \simeq 2 \times 10^7\ \mathrm{V\ m^{-1}}$, $v \simeq 10^5\ \mathrm{m\ s^{-1}}$, and $Q_\mathrm{D} \simeq 5$, therefore

$$P|Z|f^2 \simeq 10^{23}\ \mathrm{W\ \Omega\ s^{-2}}.$$

At $f = 10^{10}$ Hz and $|Z| = 10\ \Omega$ (note that this is the operating device impedance, not the low-field resistance) the maximum possible r.f. power is $P \simeq 100$ W. This numerical result may be favourably compared with those in § 7.3 where a lower and more realistic limit was placed on the r.f. voltage amplitude.

4.6. Overlength LSA oscillators

Two general criticisms can be levelled at domain-mode transit-time oscillators, even though they will operate with useful efficiencies both in theory and practice:

(1)   the Pf^2 criterion of the previous sections limits the maximum power output at the higher frequencies;

(2)   owing to the high-field concentration in the domain, it is impossible to apply too large a voltage across the device without causing impact ionization and device destruction.

A technique to avoid domain-formation and transit-time effects has already been outlined in § 3.5 and was called the LSA mode. Under these conditions the $I-V$ characteristic of the device should faithfully follow the $v-E$ characteristic. The device length is not related to the frequency and can be many times the transit-length (overlength), so that the Pf^2 criterion need not apply. Higher terminal voltages (giving larger r.f. power output) may be applied to the device without causing impact ionization. The form of the equivalent circuit for LSA operation is the same as that in Fig. 4.1, but there are several constraints to be satisfied in addition to those of § 3.5 before the mode may be achieved (Bulman *et al.* 1972)

1. $n/f \simeq 5 \times 10^{10}\,\mathrm{s\,m^{-3}}$ to satisfy the conditions for space-charge control and damping of the accumulation layer injected from the cathode (§ 3.6).
2. The internal homogeneity of the low-field conductivity must be better than about 10 percent (Copeland 1967; Hobson 1972*b*) to avoid internal injection of significant dipolar space charge (§ 3.6).
3. The device must be much longer than the space-charge transit-length per cycle, otherwise the gross inhomogeneity of the unavoidable accumulation layer will appreciably distort the uniform-field conditions.
4. The frequency-controlling cavity must be designed to allow rapid start of oscillation. Otherwise a domain mode may cause device breakdown by impact ionization before the requisite r.f. voltage amplitude has been established.

When all the above conditions are satisfied, efficiencies up to 18 percent are predicted. However, little attempt is made to achieve this single-frequency mode of operation. We will see later that the relaxation LSA mode has higher efficiency both in prediction and achievement and the above restrictions are either eased or easier to achieve.

4.7. The accumulation-layer mode

When condition 3 of the previous section is not met, the accumulation layer introduced in § 3.6 will travel an appreciable distance through the device. As was the case for domain modes discussed at the beginning of this chapter a variety of circuit-controlled modes become possible. In a resistive circuit (taken to imply a constant terminal voltage) the transit-time accumulation-layer mode (Kroemer 1966; Freeman and Hobson 1972) occurs with a much smoother waveform (Fig. 4.3) than that of a domain mode (Fig. 4.2 (a)). The accumulation layer dynamics may be understood in a simplified manner by reference to Fig. 4.3. For the present purposes the doping density is assumed to be high enough for conduction-current effects to generally dominate displacement-current effects In turn this also implies that the accumulated charge density can grow as rapidly as requested by the electric field. If a uniform electric field is applied instantaneously at time 1 an accumulation layer is injected from the cathode so that the electric-field distribution splits into two parts, as illustrated at time 2. The electron velocity on either side of the accumulation layer has been altered in the directions shown in Fig. 4.3(a). The constant terminal voltage requires that the area under each of the electric-field curves of Fig. 4.3 (b) should be equal. As the accumulation layer propagates, this equality can only be maintained if the electron velocity on either side of the accumulation layer falls as dictated by the v–E characteristic and indicated at times 3, 4, and 5. This situation continues until the accumulation layer is close to the anode and the electron velocity on

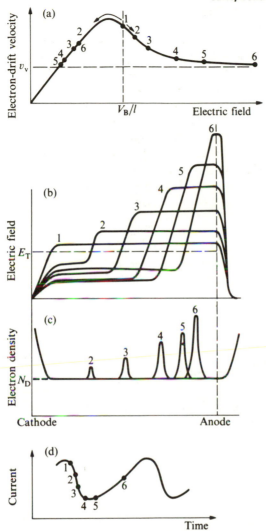

Fig. 4.3. The accumulation-layer transit mode under time-invariant terminal voltage.

either side of the accumulation layer is equal to the valley velocity. From this point the constant terminal voltage can only be maintained if the electric field on the anode side increases rapidly with time. The displacement current in this region then becomes significant and current continuity is maintained with an increased electron velocity on the cathode side of the accumulation layer, as shown at time 6. Eventually, the accumulation layer disappears into the anode, and the cathode-side field has risen through threshold so that the next accumulation layer is injected and the process repeats. The current waveform is shown in Fig. 4.3 (d).

This explanation has neglected several details such as the finite time of space-charge accumulation and the finite effects of displacement and diffusion currents throughout the cycle. It is not possible to give an analytical treatment of accumulation-layer dynamics because, in contrast to a dipole domain, it is never in a steady state. Finite-difference numerical techniques are required , but the characteristic smooth form of the current waveform is predicted in all cases where the $N_D l$ product is not large enough to allow conversion of the accumulation layer to a dipolar layer by the device non-uniformities.

When cavity control is introduced the waveforms show many similarities to domain-mode behaviour (Freeman and Hobson 1973). A delayed-accumulation mode of operation is illustrated in Fig. 4.4. A waveform similar to that of the

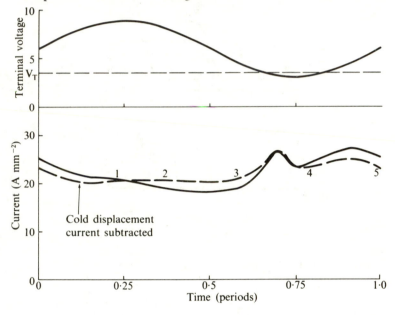

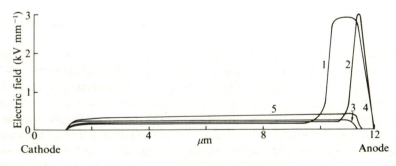

Fig. 4.4. Circuit control of the accumulation-layer mode.

delayed-domain mode arises from the similarity of the electric-field distributions at critical points of a cycle of operation. When a dipole layer or an accumulation layer are nucleated as the terminal voltage rises through threshold the electric field is initially uniform. Also, as both types of space charge enter the anode, the final phase in the process is the disappearance of an accumulation layer.

The d.c. to r.f. conversion efficiency of cavity-controlled accumulation-layer oscillators is similar to that of domain oscillators, but the best frequency of operation in devices of the same length is somewhat higher for the accumulation-layer oscillator, owing to their generally higher space-charge propagation velocity.

4.8. The hybrid mode

When n/f is a little larger ($\sim 10^{11}$ s m^{-3}) than required for LSA space-charge control or the doping uniformity does not satisfy condition 2 in § 4.6, the space-charge growth during one cycle, when the terminal voltage exceeds threshold, is sufficient for dipolar domains to form incipiently. The growth does not proceed to the quasi-static domain equilibrium inherent in the quench mode, and operation has partly an LSA and partly a quenched-domain character (Huang and Mackenzie 1968; Bott and Fawcett 1968a). As the terminal voltage falls below threshold dipolar space-charge begins to dissipate, but it is not completely quenched before the voltage again rises through threshold. Accordingly, it again grows into an incipient domain during the next cycle of operation. If the device is sufficiently long it will contain multiple domains, each of the same size but separated spatially by the transit-distance per cycle (Fig. 4.5). The dipolar space charge is eventually quenched as it enters the anode. Just after the terminal voltage has risen above threshold the electric field between the domains may also exceed the threshold field for an appreciable part of the cycle.

During the time in which the terminal voltage is just above threshold the presence of the dipolar domains causes the current to be less than it would be in LSA operation. Accordingly, the r.f. current waveform has a smaller current in antiphase with the r.f. voltage, and the efficiency of hybrid operation is less than that of LSA operation but greater than that of quench-mode devices. A comparison of the waveforms in each of the three modes is given in Fig. 4.6.

Space-charge control is easier with respect to doping homogeneity in hybrid operation than in LSA operation, so it should be somewhat easier to achieve. An advantage of the hybrid-mode over quench-mode behaviour lies in the multiplicity of domains. Their peak field is less than that of the single domain in the quench mode, so that the likelihood of avalanche breakdown is less — or alternatively the device can be operated at higher bias voltages to generate more power. Further details are given elsewhere (Bulman et al. 1972), but it should be remarked that the externally observed operating conditions are easily confused with those of the relaxation LSA mode described in § 4.10.

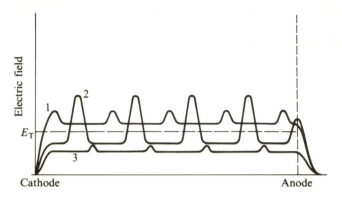

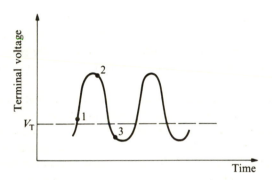

Fig. 4.5. Multiple-domain operation in the hybrid mode at several times in an r.f. cycle.

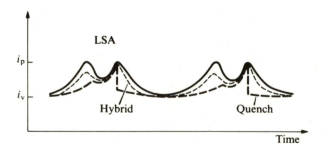

Fig. 4.6. Comparison of the current waveforms in the LSA, the hybrid and, quench modes of operation.

4.9. Calculation of efficiency and computer simulations

The efficiency of any of the modes of operation with a sinusoidal voltage wave-form at the device terminals may be calculated once the current waveform $i(t)$ is known. The r.f. power output P is given by

$$P = \frac{V_0}{2} \cdot \frac{2}{T_R} \int_0^{T_R} i(t) \cos\left(\frac{2\pi t}{T_R}\right) dt . \qquad (4.10)$$

The integral provides the current amplitude in antiphase with the r.f. voltage $V_0 \cos(2\pi t/T_R)$. Similarly, the d.c. power dissipation in the device is

$$P_{dc} = V_B \cdot 1/T_R \int_0^{T_R} i(t) \, dt. \qquad (4.11)$$

The r.f. efficiency η is given by

$$\eta = P/P_{dc}. \qquad (4.12)$$

Analytical approximations may be developed for the various modes of operation to give an explicit form for $i(t)$. For details the reader is referred to the references given for each mode.

The maximum efficiency achievable without particular reference to any mode may be derived in a simple way, without becoming heavily involved in Fourier analysis. The philosophy behind this recognizes that the maximum possible peak-to-peak current is $(I_p - I_v)$, where I_p is the current corresponding to the peak of the $v-E$ characteristic and I_v is that corresponding to the valley. The maximum current amplitude at the fundamental frequency occurs for a square waveform with an equal mark–space ratio. The r.m.s amplitude is $(\sqrt{2}/\pi)$ $(I_p - I_v)$. In order to obtain the maximum efficiency we require the instantaneous voltage to be as high as possible when the current is low, and vice versa. The minimum voltage must be the threshold voltage V_T corresponding to I_p. If the peak voltage is V_M the r.f. power generated at the fundamental frequency is

$$P = \frac{\sqrt{2}}{\pi}(I_p - I_v) \cdot \frac{\sqrt{2}}{\pi}(V_M - V_T) . \qquad (4.13)$$

The waveforms are illustrated in Fig. 4.7. The d.c. power dissipation is

$$P_{dc} = \tfrac{1}{2}(I_p + I_v) \cdot \tfrac{1}{2}(V_M + V_T), \qquad (4.14)$$

therefore

$$\eta = \frac{P}{P_{dc}} = \frac{8}{\pi^2} \cdot \frac{(I_p - I_v)(V_M - V_T)}{(I_p + I_v)(V_M + V_T)} . \qquad (4.15)$$

Cavity-controlled oscillators

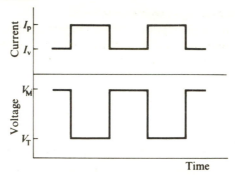

Fig. 4.7. Idealized square-wave oscillation to achieve maximum efficiency.

As the bias voltage $\frac{1}{2}(V_M + V_T)$ is raised to a high value the efficiency is asymptotically given by

$$\eta_{max} = \frac{8}{\pi^2}\left(\frac{I_p - I_v}{I_p + I_v}\right). \tag{4.16}$$

If $I_p = 2I_v$, $\eta_{max} = 27$ percent. Eqn (4.16) clearly illustrates the desirability of a material with a large peak–valley ratio.

Derivation of an analytical expression for the current waveform $i(t)$ often involves drastic assumptions which have poor justification for many practical devices. This problem can only be overcome by finite-difference computer

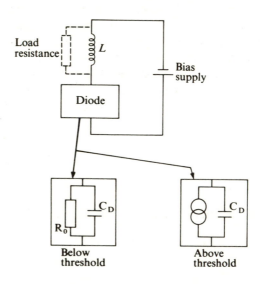

Fig. 4.8. Simplified equivalent circuit of the relaxation oscillator.

techniques to accurately calculate the time evolution of the space-charge behaviour
under the inevitable highly non-linear conditions in the device. The device under
consideration is divided into cells of finite width in space. The charge and electric
field are regarded as constant within a cell and are evaluated at regular time inter-
vals. If the cells in time and space are small enough, the solution will be a good
approximation to the exact solution. Further details are given elsewhere (Bulman
et al. 1972).

4.10. Relaxation modes of overlength oscillators

The relatively high efficiency of a device in a circuit which is suitably tuned at
its harmonics to give square-wave operation has been seen above. Unfortunately,
this waveform is not easy to achieve in practice, but the simple circuit of Fig. 4.8
allows the same objective to be acheived in a different way. A relaxation mode
of operation may be set up which essentially achieves the desired objective of
switching rapidly between a low-current high-voltage condition and its converse.
The voltage and current waveforms are illustrated in Fig. 4.9. When the terminal
voltage is rising and below threshold, the diode is ohmic, with a relatively-small
parallel capacitive susceptance. The current is exponentially limited by the L/R_0
time constant. When the voltage exceeds threshold, the current begins to fall
rapidly and the diode behaviour is similar to that of a current source (the valley
current) in parallel with the capacitive susceptance. The rapid fall of current
through the inductance causes a sharp rise of terminal voltage, which executes
an incomplete cycle with the natural period of the inductance and device capacity.

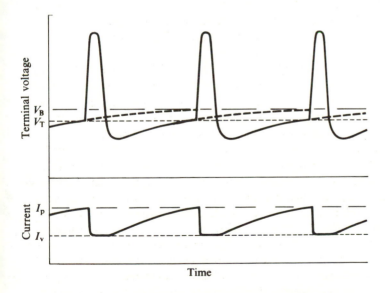

Fig. 4.9. Current and voltage waveforms of the relaxation oscillator.

Cavity-controlled oscillators

Space charge is quenched when the voltage swings below threshold, and the cycle repeats. Heavy loading of the oscillator causes the 'ringing' part-cycle to approach a complete natural period, but it is reduced towards a half period by the larger amplitude accompanying lighter loading (Jeppesen and Jeppsson 1971; Wasse, Mun, and Heeks 1972). Periodic oscillation will not occur if the circuit loading is too heavy, because the transient voltage will not swing below threshold.

Increasing the bias voltage causes a reduction of the time spent below threshold, so there is a characteristically large increase of frequency with bias voltage (see § 8.7) which is a signature of the mode. Care has to be taken in this interpretation owing to similar but weaker behaviour by other sinusoidal overlength modes. Optimum operation requires that the asymptotic frequency f_∞ (obtained as the bias voltage tends to infinity) be related to the electron density by (Camp, Eastman, Bravman, and Woodard 1972)

$$n/f_\infty \simeq 1\cdot5 \times 10^{11}\,\mathrm{s\,m^{-3}}. \tag{4.17}$$

This requirement is essentially an LSA condition relaxed by the rapid transition of electric field through the maximum negative-differential mobility of the $v{-}E$ characteristic. The rapid-switching action also makes the relaxation oscillator less susceptible to doping non-uniformities than the sinusoidal LSA oscillator (Jeppesen and Jeppsson 1971). Eqn (4.17) ensures a small enough device susceptance so that the current change, as the voltage rises through threshold, is not diverted from the inductance into the parallel capacity. Otherwise, sinusoidal operation will occur at a frequency controlled by the inductance and capacitance, and operation will be in one of the modes described at the beginning of this chapter. In a similar way spurious parallel capacity (from packages, etc.) can also preclude relaxation oscillation (Solomon, Shaw, and Grubin 1972).

Simple circuits have evolved (Fig. 4.10) (Eastman 1972; Wasse *et al.* 1972), after the expense of much pain and time, to provide nearly instantaneous feedback of voltage by the inductance at the instant of threshold crossing. It is not sufficient to have a transmission-line equivalent inductance, since the signal delay generally prevents simultaneous current fall and voltage overswing.

When the above conditions are achieved the efficiency of relaxation oscillators can approach 28 percent but ~15 percent is typical. Experimentally, it is difficult to determine the mode of operation conclusively. The relaxation operation can also occur for domain modes or hybrid modes. The most direct evidence for an LSA form, or , at most, a multiple-domain distribution (Solomon *et al.* 1972), arises from the avalanche-breakdown behaviour. This will often occur when a relaxation oscillator is heavily loaded so that transit-frequency domain operation is enforced (at a much lower frequency). Breakdown in the same device may be prevented in high-efficiency operation at the same bias voltage when the circuit is correctly adjusted. It is inferred that the proper oscillator operation prevents formation of high-field, mature domains.

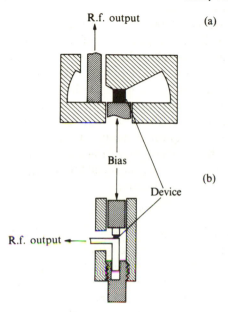

Fig. 4.10. Typical LSA-relaxation circuits: (a) Eastman (1972), (b) Wasse, Mun, and Heeks (1972).

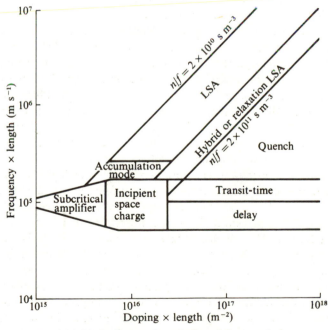

Fig. 4.11. Mode diagram for transferred-electron devices.

Cavity-controlled oscillators

4.11. A mode diagram

The relationship between the doping, frequency, and device length for the various transferred-electron devices is illustrated in Fig. 4.11. The boundaries between each type of behaviour should not be regarded as absolute. They have a significant dependence on the bias voltage, the conductivity uniformity, the device temperature, and the circuit loading. For example, the transit-time mode only exists in resistive circuits, and the appropriate circuits will cause the quench mode and delay mode to merge and overlap. If the doping homogeneity is poor, the LSA mode may be non-existent or near steady-state domains may exist in the region of incipient space-charge modes. Analytical approximations for calculation of the operating parameters can only be made for doping and frequency parameters around the edge of the mode diagram.

4.12. Incipient space-charge modes

Practical c.w. devices are made in a longitudinal geometry and must have as short a length as possible to ease problems of heat-sinking (see § 6.3). The $N_D l$ product must not be too small, otherwise the efficiency and r.f. power output will be too small. Compromise between these two requirements to give maximum r.f. power usually requires $N_D l \simeq 1 \times 10^{16} \,\mathrm{m}^{-3}$ to $2 \times 10^{16} \,\mathrm{m}^{-3}$, and the mode of operation is a mixture of all the different types of mode. The relative proportions depend quite sensitively on the bias voltage, the conductivity uniformity, and the circuit loading. Finite-difference computer simulations are required for satisfactory calculation of their behaviour (Freeman and Hobson 1972, 1973). Some properties of these devices are significantly different from the 'classical' Gunn devices with much larger $N_D l$ products. The c.w. transit-frequency of devices mounted in resistive circuits is strongly dependent on the bias voltage, particularly for devices with alloyed-cathode contacts (Hobson 1967). The product of bias voltage and transit-frequency (the Vf_T product) is approximately constant in the range 100 GHz V to 150 GHz V, whereas the 'classical' Gunn diode has a transit-frequency which is substantially independent of the bias voltage. A related phenomenon is the requirement for an optimum bias voltage to give optimum r.f. power output in cavity-controlled operation. The optimum bias voltage depends on the circuit-controlled frequency, and again there is a nearly-constant product of bias voltage and frequency (the $V_B f$ product) to give optimum performance. In this case the product is usually 25 percent to 35 percent lower than that for transit-frequency operation (Hobson 1967). Similar behaviour is observed with devices whose cathode contact is made of n$^+$ GaAs, but the voltage sensitivity lies between that of the devices with an alloy cathode contact and that of 'classical' devices with a large $N_D l$ product. In pulsed operation the optimum product of bias voltage and frequency is not observed for either type of device, and operation at frequencies giving the best efficiencies results in a monotonic increase of power and efficiency with increasing bias voltage. The monotonic increase cannot be explained by analytical theories of circuit control involving stable domain propagation (Warner 1966). For these devices, satisfactory calculations can only be made with a finite-difference computer simulation, owing to the near equality of the space-charge growth and transit-times. Simulations indicate that the differences between the devices with an alloyed cathode contact and those with an n^+ cathode contact originate from a cathode-doping defect in the former device, which injects a large quantity of dipolar space charge (Freeman and Hobson 1973). The $V_B f$ and $V f_T$ products arise from a combination of thermal and space-charge growth effects, which cause a decrease of the mean space-charge velocity with increasing bias voltage (Freeman and Hobson 1972), so that optimum space-charge synchronization with the r.f. terminal voltage occurs at a lower frequency. The computer simulations also exhibit the monotonic increase of r.f. power and efficiency of the same devices in pulsed operation where bias-dependent heating effects are absent.

4.13. Energy-relaxation effects

The effects of energy relaxation on the electron energy distribution have already been introduced. They set an upper frequency-limit between 50 GHz and 100 GHz when uniform field conditions are assumed to prevail and cause diffusion-like effects in stable, but non-uniform, space-charge profiles. Energy relaxation will also modify the time-varying behaviour in realistic devices with non-uniform electric-field profiles. Jones and Rees (1972) have identified a spatial delay of about $2 \, \mu$m in the nucleation of space-charge from doping non-uniformities, owing to the finite time required for electrons to achieve enough energy for intervalley transfer. In addition, any modes which require the quenching of space charge suffer an upper frequency-limitation of approximately 20 GHz, owing to the time (~ 20 ps) required for electrons in the satellite valleys to transfer to the central valley and lose energy (thermalize). If the terminal voltage rises above threshold again in a shorter time, gradients of electron temperature (i.e. mobility) will still exist in the quenching region and act as a space-charge nucleating non-uniformity. Accordingly the correct phasing of the space-charge transit with the terminal voltage will be destroyed, and cyclic operation will not be possible. Jones and Rees (1972) suggest that the highest frequencies may be achieved with an accumulation-layer transit mode, because all the electrons which have to thermalize at the low-voltage part of a cycle do so in the heavily doped, anode region. They further suggest that dipolar domain operation will not occur at the higher frequencies owing to the shortage of time for the depletion layer to form. At the time of writing a comprehensive survey of these effects has not been carried out. The explanation of many detailed features of X-band oscillators by the effects of transient space-charge behaviour (Freeman and Hobson 1972, 1973) suggests that energy-relaxation effects will only be dominant at higher frequencies.

47

5.1. Introduction

Several derivatives of the transferred-electron effect have been used as the basis for amplification. They can be divided into two classes. One utilizes devices with internal stability while the second one uses c.w. Gunn oscillators to provide amplification from the negative-differential conductance in their bias circuit and from parametric effects caused by non-linearity in their susceptance or conductance.

5.2. Subcritical amplifiers

Devices with an $N_D l$ product less than $5 \times 10^{15}\,\mathrm{m}^{-2}$ do not support internal space-charge instability. It may be thought that this would allow the device to be used as a negative resistor, but Nature dictates otherwise. In subcritical devices with ohmic contacts a highly non-uniform but static distribution of electric field and space charge is set up when the mean field exceeds threshold. We will see that this implies a monotonic increase of current with terminal voltage providing diffusion effects are negligible (Shockley 1954). Detailed discussions are available elsewhere (McCumber and Chynoweth 1966; Mahrous and Robson 1966; Holstrom 1967; Bulman *et al.* 1972), and only a plausible interpretation of the results will be given here. The reader should note that there is a convention, often used in problems involving electrons alone, in which the electron charge is regarded as positive. It will be used here. The non-uniform field and space-charge distributions are illustrated in Fig. 5.1. They are necessary to satisfy the $v-E$ characteristic, Poisson's equation

$$\frac{\partial E}{\partial x} = \frac{e(n - N_D)}{\epsilon \epsilon_0},$$

and current continuity

$$J = nev(E) + \epsilon \epsilon_0 \frac{\partial E}{\partial t} + eD \frac{\partial n}{\partial x}.$$

For the moment diffusion effects are neglected and recognition of the static nature of the problem gives

$$J \simeq nev(E) = \text{constant}.$$

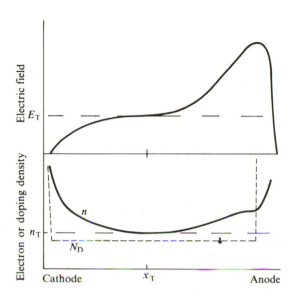

Fig. 5.1. The non-uniform distribution of electric field and electron density in a subcritically doped amplifier. n is the electron density and N_D is the donor-doping density.

Close to the cathode the electric field is of necessity small, so there must be an excess electron density in order to carry the current required by current continuity and to allow the positive $\partial E/\partial x$. Progressing into the device the field rises, causing $v(E)$ to rise while $\partial E/\partial x$ and the accumulated charge decrease, still maintaining current continuity, until the electric field approaches threshold at x_T. This is the point of highest electron-drift velocity and therefore of lowest electron density and smallest $\partial E/\partial x$. On the anode side of x_T the electric field exceeds threshold, the electron velocity falls, and $\partial E/\partial x$ and the accumulated space-charge increase. This trend continues up to the anode, where the steep increase of donor density exceeds the electron density and the electric field falls to zero. The electric field penetrates into the anode (typically by 1 μm in material doped to 10^{21} m^{-3}), where the integrated depletion of charge is equal to the integrated accumulation in the bulk of the device.

Unfortunately, the terminal conductance of the diode is not negative under these non-uniform conditions. An increase of d.c. voltage requires an increase of the integral of electric field over the full length of the device, so that there is an increase of accumulated charge to support the increased gradients of the electric field. Under these conditions the current increases monotonically with terminal voltage. This conclusion may be seen most clearly at the point x_T, where $\partial n/\partial x$ is zero so that the diffusion current is zero and the total current is $i_B = n_T e v_T$. Increase of $\partial E/\partial x$ at x_T with increasing terminal voltage causes n_T and therefore i_B to increase. The current—voltage characteristic is shown in Fig. 5.2.

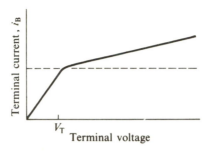

Fig. 5.2. A typical current–voltage characteristic for a subcritically doped amplifier.

Despite the above positive-conductance character Nature may be outflanked under some conditions, and a terminal negative conductance can exist over a restricted frequency range. When the bias is applied the accumulated charge flows into the device from the cathode, where it is replaced, through the external circuit, by the depleted charge in the anode region. This process takes a finite time within the device and allows a negative conductance to occur, as may be seen by considering the response of a device to a step function change of current. If the device was previously in equilibrium, as illustrated in Fig. 5.1, the initial current change will be carried as displacement current and the field will increase with time throughout the device. x_T will move towards the cathode, and there will be a velocity 'redistribution' followed by a space–charge redistribution. The electron-drift velocity is highest at x_T and decreases montonically in the higher-field regions towards the anode, so that there is a transient bunching effect towards the anode on the anode side of x_T. Consequently, a sympathetic movement of electric field, necessary to satisfy Poisson's equation, will cause its integral from cathode to anode to fall to a value somewhat less than the initial steady-state terminal voltage. A transient, negative-differential conductance exists. These events occupy a time comparable with the space-charge transit-time. At this time, the displacement current in the cathode region will be decaying and more space-charge is injected from the cathode to maintain current continuity. When it has travelled through the device and steady conditions have been achieved again, the field is higher close to the anode and it has an integral greater than the terminal voltage before application of the current step. Fourier transformation of this temporal current–voltage behaviour into the frequency domain shows that a negative-differential conductance will exist at frequencies close to the inverse of the transit-time (Fig. 5.3). Some negative conductance may be expected at harmonics of this frequency, but the effects are damped out by diffusion at the higher frequencies (Kroemer 1967). Comments on the value of subcritically doped amplifiers will be made at the end of the next section.

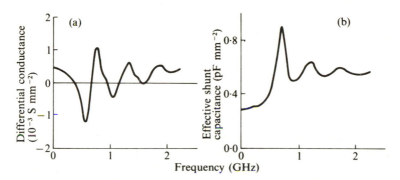

Fig. 5.3. Differential conductance and capacity of a subcritical amplifier. Length = 200 μm, $N_D = 10^{19}\,\text{m}^{-3}$, mean bias field = 475 V mm^{-1} (McCumber and Chynoweth 1966).

5.3. Reflection-amplifier circuits

Negative-conductance devices at microwave frequencies are usually mounted at the end of a transmission line or waveguide to give reflection amplification as illustrated in Fig. 5.4. The power gain g on reflection is given by (Montgomery, Dicke, and Purcell 1948)

$$g = \left| \frac{Y_0 - Y(\omega)}{Y_0 + Y(\omega)} \right|^2 , \tag{5.1}$$

where Y_0 is the characteristic admittance of the transmission line and $Y(\omega)$ is the admittance of the diode and its connecting circuit. If $Y(\omega)$ was purely conductive and negative, g would have a frequency-dependence directly related to the curve shown in Fig. 5.3(a). In practice there is always a finite capacitive susceptance in parallel with the negative conductance, and it must be tuned out by

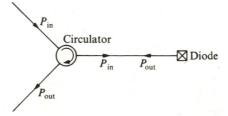

Fig. 5.4. Schematic diagram of a reflection amplifier. The circulator is used to separate input and output signals.

an inductive element at the required centre frequency. If this element is a lumped parallel inductance it may be shown that the relationship between the gain at the

Amplifiers

centre frequency g_c and the 3 dB bandwidth B is (see Appendix II)

$$\left(g_c^{\frac{1}{2}} - 1\right)\left(1 - \frac{2}{g_c}\right)^{\frac{1}{2}} B = \frac{G}{\pi C}. \tag{5.2}$$

If $g_c \gg 1$ eqn (5.1) simplifies to

$$g_c^{\frac{1}{2}} B = \frac{G}{\pi C}. \tag{5.3}$$

G and C are the negative conductance and capacity respectively and have been assumed to be frequency-independent over the frequency range B. $g_c^{\frac{1}{2}}$ in the above equations is equal to the voltage gain. $g_c^{\frac{1}{2}} B$ is usually referred to as the voltage-gain—bandwidth product and is a useful figure of merit.

The performance of subcritical amplifiers is not good enough to give them any systems use. They have poor voltage-gain—bandwidth products (100 MHz is typical) and low saturation-power output (between 1 mW and 100 mW), and their noise figure (20 dB to 25 dB) is too poor for them to be useful as low-level devices. The lack in the first two properties is largely a result of the scarcity of electrons in these devices. Encouraging improvements are available from the so-called 'supercritical' devices discussed next.

5.4. Supercritical amplifiers

In the early days of Gunn devices it was presumed that so-called supercritical devices ($N_D l < 5 \times 10^{15}\,\mathrm{m}^{-3}$) were inherently unstable under all conditions, so little attempt was made to develop them as amplifiers. However, devices with $N_D l \quad 3 \times 10^{16}\,\mathrm{m}^{-3}$ may be stabilized (Perlman, Upadhyayula, and Siekanowicz 1971) if care is taken to avoid resonances of the device with its circuit at those frequencies where the negative conductance is greater than the load conductance. Clean microwave circuits are required with avoidance of spurious resonances at frequencies other than those desired, and the conductive loading of the circuit (Y_0 in eqn (5.1)) must be greater than the modulus of the negative conductance. A simple method of stabilization is to increase the iris coupling of a waveguide tuned circuit with a file or to increase the magnetic-loop coupling to a coaxial cavity until self-oscillation ceases. Stabilization becomes easier both for high bias voltage and for c.w. operation (Perlman, Upadhyayula, and Siekanowicz 1971; Charlton and Hobson 1973) but stable amplification is also possible in some devices biased just above the bias-current saturation point (Kennedy 1966). In both cases the negative-differential conductance of the v–E characteristic at the mean bias field is less than the maximum, so that the device is not such a powerful negative resistor. Voltage-gain—bandwidth products in X-band (8·2 GHz to 12·4 GHz) have a best performance of 15 GHz, but 5 GHz is more typical and imperfect circuit design (essentially a large spurious capacity in parallel with the device) will often reduce this to ~1 GHz. In Q-band (26 GHz to 40 GHz) the voltage-gain—bandwidth products are limited to $\sim 1\cdot5$ GHz (Baskaran and Robson 1972a), owing to extreme difficulty in the design of 'clean' circuits and avoidance of parasitic capacity and inductance around the device.

For an amplifier to be useful in a microwave receiver it must not add a significant amount of noise to the desired signals. This property is often described by the noise figure which is (signal-to-noise ratio at input/signal-to-noise ratio at output). Unfortunately, transferred-electron devices do not perform well in this respect. The noise figures of supercritical amplifiers are substantially independent of centre frequency between 6 GHz and 40 GHz. Typically they lie between 15 dB and 20 dB in GaAs, with a best performance of ~ 10 dB. Indium-phosphide supercritical amplifiers have a better noise performance, with typical noise figures of ~ 12 dB and a best performance of $7\cdot5$ dB (c.w.) in Q-band (Baskaran and Robson 1972b) and $8\cdot8$ dB (pulsed) in X-band (Braddock and Gray 1973). The reasons for the better noise performance in InP are not known, but it may be an indication of the different ratios of diffusion constant to mobility in the two materials (Thim 1971). This ratio is essentially a measure of the mean energy of the random distribution of electron energies, so the implication is one of less voilent electronic motion in InP.

If the amplifiers are not particularly promising at low levels we must look for their desirable features as higher-power amplifiers. Requirements exist at ~ 1 W output both for amplifiers to drive travelling-wave tubes and for lower-power-output amplifiers in some communication systems, so attention is now focused on the non-linear properties of transferred-electron amplifiers. The incident power level at which gain saturation occurs is determined by the r.f. voltage amplitude at the diode terminals, because this quantity controls non-linear modifications of G at large signal levels. A typical relationship between G and the r.f. voltage amplitude V_0 is shown in Fig. 5.5. Fig. 5.6.shows a typical relationship between input and output power for various circuit adjustments. The input power level at which appreciable non-linearity occurs clearly decreases with increasing gain, but the output power saturation is substantially unaffected and is approximately equal to the maximum oscillator power under the same bias conditions. As shown in Appendix III the above recognition that non-linearity is determined by the r.f. voltage at the diode terminals implies that the maximum r.f. power $(P_{net})_{max}$ that the device can *add* to the input power is equal to the maximum oscillator power under the same bias conditions. The centre-frequency gain is related to the input power $(P_{in})_n$ for a given degree of non-linearity by

$$g_c - 1 = (P_{net})_{max}/(P_{in})_n . \qquad (5.4)$$

This relationship is illustrated in Fig. 5.7. When $g_c \gg 1$,

$$g_c(P_{in})_n \simeq (P_{net})_{max} = \text{constant}. \qquad (5.5)$$

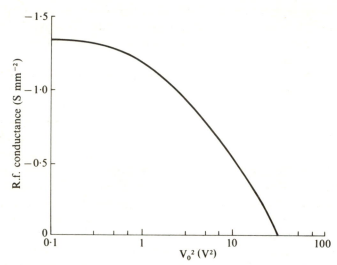

Fig. 5.5. A typical variation of r.f. conductance with r.f. voltage amplitude for an X-band supercritical amplifier.

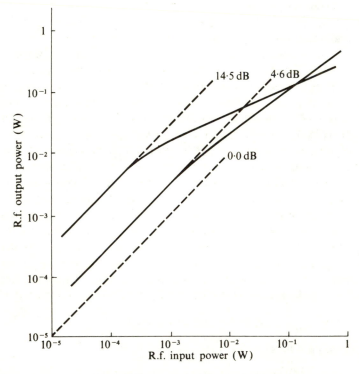

Fig. 5.6. Typical gain-saturation behaviour of an X-band supercritical amplifier at $10 \cdot 5$ V bias. The -1 dB gain-compression point occurs at approximately the same power output in each case at ~ 15 mW.

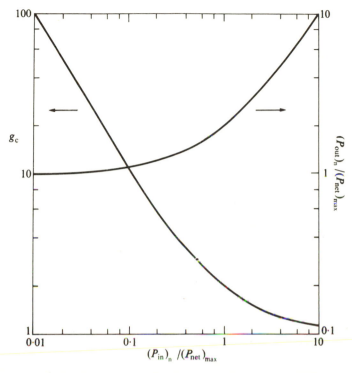

g_c

$(P_{out})_n / (P_{net})_{max}$

$(P_{in})_n / (P_{net})_{max}$

Fig. 5.7. The relationship between the input power, at which non-linearity is evident, and
(a) the amplifier gain or (b) the output power at this point of non-linearity.

Eqn (5.4) indicates that decoupling of the diode (e.g. by off-centre waveguide
mounting) allows maintenance of usable gain to incident and reflected power
levels well in excess of the maximum oscillator power under the same conditions
(Fig. 5.8). The diode adds power up to the maximum oscillator power. The disadvantage is low gain. This added-power feature is useful for cascaded-amplifier
design, where high-gain, low-saturation adjustment may be made for the input
stage and vice versa for the output stage.

At the time of writing the stabilization mechanism is the subject of much
debate. Small-signal calculations show the possibility of increasing the critical
$N_D l$ product when the r.f. short-circuit boundary condition is replaced with a
finite admittance in the external circuit (Sterzer 1969), but there has not been
a comparable large-signal calculation to demonstrate circuit-induced stabilization
of large-signal space charge injected from the contacts.

A large-signal stabilization mechanism proposed by Charlton, Freeman, and
Hobson (1971) essentially involves raising the critical $N_D l$ product by conduc-
tivity-profile modifications. When a device has a uniform doping profile and
ohmic contacts the electric-field distribution above threshold has the form shown

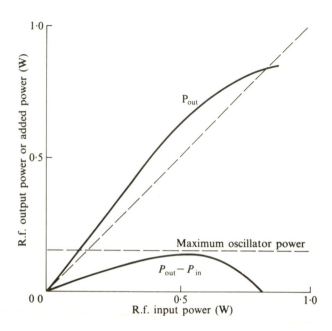

Fig. 5.8. Addition of power by a supercritical amplifier at incident- and reflected-power levels well in excess of the oscillator power.

in Fig. 5.1. The region close to the threshold field (where the negative-differential conductivity is greater) is the most critical for formation of instabilities. The doping profile illustrated in Fig. 5.9 is designed to make the electric field pass through threshold in a short distance and to maintain a near-uniform field well above threshold. Under these circumstances the critical $N_D l$ product may be raised to approximately 3×10^{16} m^{-3}, because the average negative-differential mobility of the v–E characteristic is small at high fields. A device with uniform doping, other than in the 'notch' region at the cathode, is considered first. The steep rise of electric field from the cathode is caused by the excess of electrons (required for current continuity) over donors in the notch. If the bias voltage is just the correct magnitude, the electric field in the rest of the device is substantially uniform up to the anode, as required by current continuity. The required bias voltage is equal to the product of the device length and the field achieved on the anode side of the doping notch and is somewhat inflexible. It is also much greater than threshold if the uniform field is to be substantially higher than that required for the steepest negative-differential conductivity, as it must be if the critical $N_D l$ product is to be raised appreciably. The bias-voltage range for stabilization can be increased (towards lower voltages) by introducing a doping profile with an upward slope from cathode to anode, as shown in Fig. 5.9. As the bias voltage is reduced the nearly-constant field remains almost unaltered except

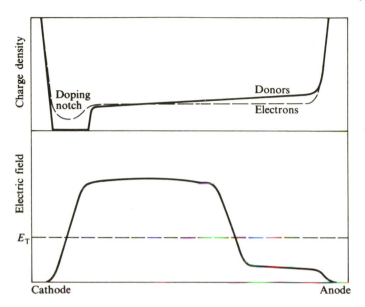

Fig. 5.9. Electric-field and space-charge profiles for 'cathode-notch' stabilization of super-critical amplifiers.

that its anode-side steep slope moves towards the cathode. The doping slope acts as a quasi-anode and allows the steep field gradient to exist in a voltage-dependent position outside the anode contact (Charlton, Freeman, and Hobson 1971). Stabilization at the lower bias voltages is aided by the effective reduction of device length as the bias voltage is decreased. Broad agreement is found in the comparison of this theoretical approach with the practical details described earlier (Charlton and Hobson 1973). Similar field profiles can be generated by notch reduction of the electron mobility at the cathode and by barrier profiles.

An alternative mechanism requires the electron density to be high enough to allow the flow of significant diffusion currents. The cathode-side field lies below threshold but rises steeply close to the anode (Fig, 5.10) (Magarshak and Mircea 1970; Jeppesen and Jeppsson 1972). A minimum electron density is required in order to allow a large enough reverse diffusion current to flow in the region where the field rises through threshold so that current continuity may be maintained with static field in this region of high electron-drift velocity. An $N_D l$ product greater than a minimum somewhere between $10^{16} \, \text{m}^{-2}$ to $3 \times 10^{16} \, \text{m}^{-2}$ is required for a doping density of about $\sim 10^{21} \, \text{m}^{-3}$.

Both models give easiest stabilization at the highest bias voltages and uncertainty about the diffusion coefficient does not allow easy decisions on the basis of the mean electron density. Even though the doping profile may be determined on test pieces of material before contacting the free surface, it is normally

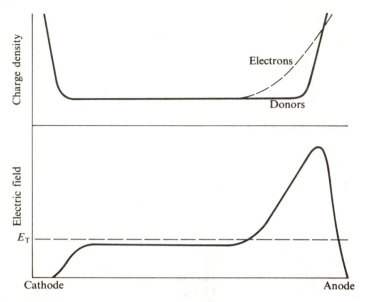

Fig. 5.10. Electric field and space-charge profiles for 'anode-field' stabilization.

impractical to determine both the doping and mobility profiles of practical devices – particularly that part beneath the contacts which may be damaged in device fabrication. However, trends in the noise performance of Q-band amplifiers with different test-piece doping profiles are not inconsistent with the cathode-notch stabilization mechanism (Baskaran and Robson 1972a). A further unknown is the effect of electron energy relaxation on steep gradients of electric field and space charge.

If all the above complexities surround stabilization of supercritical amplifiers it is reasonable to ask why they are more desirable than subcritically doped reflection amplifiers. The predominant advantage of the former is their higher power-handling capacity owing to their greater electron content for a given impedance level, which, in turn, is set by the device capacity. Practical saturation powers are ~ 1 W in X-band so the supercritical amplifiers have appeal for communication-systems use in low-power transmitters.

5.5. Travelling-wave or two-port amplifiers

Growth of space charge in a stabilized negative-conductivity medium has been exploited in the construction of two-port unilateral amplifiers (Robson, Kino, and Fay 1967). Electromagnetic signals are coupled to the device, near to the cathode, with a pair of electrodes (Fig. 5.11). The space charge which is induced travels and grows with the drifting electrons, and amplified signals are recovered

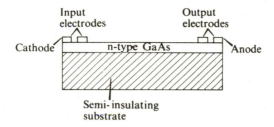

Fig. 5.11. Schematic diagram of a two-port amplifier.

by a pair of electrodes at the anode. One of the problems with this type of amplifier is matching at the input and output ports. It appears that the coupling efficiency at the cathode between the input electric field and the space charge can be improved with a Schottky barrier as the second contact, so that it modulates the electron density rather like a field-effect transistor (Kanbe, Shimizu, and Kumabe 1973). The anode contacts should be ohmic to obtain optimum coupling of space charge to the current in the metallized coupling lines. The bandwidth of these amplifiers is much greater than the two-terminal devices described above, owing to the absence of a transit-time limitation. Stabilization is often aided by making the lateral dimensions comparable with the critical $N_D d$ product, so that the critical $N_D l$ product is increased. Further developments in this direction include the use of dielectric surface loading. An extensive discussion of all these effects is given by Dean and Maturese (1972). Present results include a net gain of 28 dB at 9·2 GHz. Net gains have been observed up to 20 GHz. Unfortunately, two-port amplifiers have poor noise figures (typically $>$ 20 dB) and gain saturation at low power output (a few milliwatts), so they have not yet filled any natural requirements of microwave systems. One other possible application is for use as a phase shifter using the change of the electron space-charge drift velocity with bias voltage. The shape of the $v-E$ characteristic makes the phase shift approximately proportional to the bias voltage (Robson, Kino, and Fay 1967).

5.6. END conductance

The current in the bias circuit of a Gunn oscillator often shows a decrease with increasing bias voltage above threshold. In c.w. operation the effect is usually dominated by the bias-voltage dependence of the temperature (Freeman and Hobson 1972), but the exclusion of thermal effects by use of sufficiently rapid voltage changes will often leave a residual *external negative-differential* (END) conductance. This END conductance causes undesirable bias-circuit oscillations if care is not taken with the circuitry but has also been used as the basis of a microwave amplifier (Thim 1967; Charlton, Hobson, and Martin 1972) and u.h.f. and v.h.f. oscillators (Cawsey 1967).

Amplifiers

The simple explanation of these effects has assumed that the END conductance arises from the negative slope of the current-voltage characteristic of a transit-time Gunn diode containing a domain in flight, as shown in Fig. 5.12 (Thim 1967). However, the external current—voltage relationship of a Gunn oscillator in cavity-controlled operation is dependent on its conductive loading, and it has been shown both theoretically (Hobson 1969*b*) and experimentally (Charlton, Hobson, and Martin 1972) that a lightly-loaded cavity-controlled oscillator will give a larger END conductance, whereas heavy loading of the oscillator tends to weaken the effect.

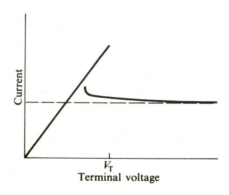

Fig. 5.12. The terminal current—voltage characteristic of a long transit-time Gunn oscillator containing a steady-state domain.

5.7. Parametric amplification

Space-charge layers or domains constitute a capacity which is time-varying in an oscillating diode both in the period of growth and at the times of injection or quenching. The self-pumped capacity appears at the device terminals, and it may form the basis of a simple parametric amplifier or oscillator. Both degenerate (Aitchison, Corbey, and Newton 1969) and non-degenerate amplifiers (Kuno 1969) have been built, and they generally have voltage-gain—bandwidth products which are a little better than the END-conductance devices mentioned previously. A gain of 30 dB with a voltage-gain—bandwidth product of 50 MHz and a minimum single-sideband noise figure of 18 dB has been measured (Kuno 1969). On the unfavourable side these effects may result in the simultaneous operation of an oscillator at three or more frequencies if the circuit design is poor. Similar properties are expected from non-linearity of the negative conductance of an oscillator (Robson and Hashizume 1970), but it is difficult to isolate the effects experimentally.

<table>
<tr><td>

6

</td><td>

Material growth, device fabrication, and thermal design

</td></tr>
</table>

6.1. Material growth

Many 'adolescent' devices have been made with dices cut from material which was grown as a bulk single crystal from the melt by the Czochralski or Bridgman methods or their variations. Such material has a large compensated impurity density which causes a low electron mobility. The electron density is strongly dependent on temperature, so that thermal runaway readily occurs under c.w. conditions. All useful devices for systems applications are made from epitaxial GaAs. Only a brief outline of device technology and of GaAs crystal growth by liquid- (or vapour-) phase epitaxy will be given here. More detailed accounts are available elsewhere (Bulman *et al.* 1972).

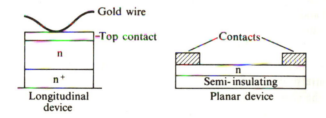

Fig. 6.1. Device types on high-conductivity or semi-insulating substrates.

Both epitaxial techniques require a substrate of bulk-grown, single-crystal GaAs, typically 20 mm in diameter and 1 mm thick, to define the single-crystal growth and to give mechanical strength to the thin active region of the device (usually $< 100 \ \mu$m). Two types of device may be identified (Fig. 6.1). The most common is the longitudinal one, where current flow is perpendicular to the substrate which must have a high conductivity. Planar devices have a current flow parallel to the substrate, which must be insulating. Many difficulties are encountered with interface defects and with surface breakdown in the latter geometry, so it finds little use at the time of writing. It has become a common practice to grow an epitaxial 'buffer' layer of highly-doped n^+ material on the highly-doped substrate (or semi-insulating material on a semi-insulating substrate) in order to reduce propagation of crystal defects and diffusion of impurities from the substrate into the active region of the device.

61

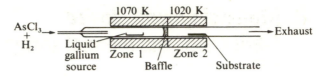

Fig. 6.2. Schematic diagram of a furnace for epitaxial growth from the vapour phase.

Vapour-phase epitaxial growth requires compounds which may be mixed as vapours and combine to give GaAs at the substrate surface. Availability of high-purity materials usually restricts the choice to the arsenic trichloride ($AsCl_3$) system or the arsine (AsH_3) system. The arsenic-trichloride system is sketched in Fig. 6.2 and the arsine system is similar. High-purity hydrogen is bubbled through $AsCl_3$ which is carried into the furnace, where the high temperature causes its reduction to As and HCl vapour,

$$4AsCl_3 + 6H_2 \rightarrow As_4 + 12HCl .$$

Arsenic reacts with the high-purity molten Ga until saturation is achieved with a skin of GaAs on its surface. At this point in time a substrate is introduced into the second zone of the furnace. GaCl (produced by the reaction of HCl with Ga in zone 1) and As_4 vapour combine in zone 2 at 1020 K, and epitaxial crystal growth occurs on the GaAs substrate according to the reaction

$$As_4 + 6GaCl \rightarrow 4GaAs + 2GaCl_3 .$$

If the substrate is placed in a higher-temperature region of the furnace just after insertion, the above reaction is reversed, and the surface will be etched and surface contamination removed before it is moved to the growth region at a lower temperature. n-type doping of the epitaxial layers may be achieved with sulphur or selenium, which may be introduced as a controlled flow of hydrogen sulphide or hydrogen selenide. Alternatively, dopants can be introduced in the molten Ga.

Vapour-phase epitaxy is capable of producing material with a background doping level of $10^{20}\,m^{-3}$. The electron mobility at 77 K (commonly used as an indicator of material perfection) may exceed 10 $m^2\,V^{-1}\,s^{-1}$, while at 300 K it may be 0·9 $m^2\,V^{-1}\,s^{-1}$. One of the great advantages of vapour-phase epitaxy is its ability to provide very thin layers down to 1 μm (at a typical rate of 5 μm to 15 $\mu m\,h^{-1}$). Lengths up to 100 μm may be produced, but liquid-phase epitaxy is often more successful at this extreme.

Liquid-phase epitaxial growth is achieved by crystallization of GaAs from a suitable solution in contact with the substrate. A typical solution composition is 10 atom − % As in 90 atom − % molten Ga. Crystallization is induced by cooling and the particular nature of the Ga−As−GaAs phase diagram (Bulman *et al.* 1972) causes deposition of GaAs at the appropriate temperatures. High-purity material results from this process because of preferential segregation of impurities into the liquid phase.

Residual impurity densities from liquid-phase epitaxy may be as low as 10^{18} m^{-3}, and the electron mobility at 77 K can be 33 $m^2 V^{-1} s^{-1}$. The room-temperature mobility may closely approach the theoretical limit slightly in excess of $0.9 \, m^2 V^{-1} s^{-1}$. In order to produce the desired electron density it is necessary to introduce suitable amounts of the appropriate impurity into the molten solution. Sn, Se, or Te are commonly used dopants. Care has to be taken with the deposition temperature of GaAs with Sn dopant in order to avoid doping compensation or the production of p-type material, because Sn is an amphoteric dopant (i.e. it can give n-type material when it enters the crystal structure on a Ga site or p-type material when it enters on an As site). Good doping uniformity may be obtained over lengths of 100 μm making liquid-phase epitaxy eminently suitable for producing overlength LSA-relaxation oscillators or similar devices.

6.2. Diode construction and contacts

Individual diodes are made from the epitaxial layers either by depositing many contacts on the exposed epitaxial surface followed by dicing the layer into individual units typically 0.5 mm square or vice versa. Electrical contact to the diode environment usually requires low-resistance ohmic contacts, even though it may be advantageous for some devices to have a controlled-barrier contact at the cathode (Yu, Tantraporn, and Young 1971). Ohmic contacts are made with alloyed metals or highly-doped n^{+} regrown epitaxial layers. One ohmic contact will have been made (hopefully) to the highly-doped substrate of a longitudinal device during the epitaxial growth process. It is a relatively simple procedure to 'solder' the highly-doped region onto a metal support.

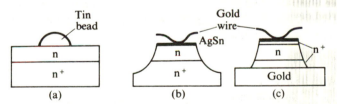

Fig. 6.3. Longitudinal device structures: (a) tin bead, (b) mesa, and (c) inverted mesa.

In the infancy of a particular device-type the bead contact (Fig. 6.3 (a)) is often used for simplicity. A bead of Sn, AgSn, or AuGe is placed on the exposed epitaxial surface of the already-diced device. The bead is alloyed into the GaAs surface at an elevated temperature in an inert atmosphere with the addition of HCl vapour to promote surface wetting and cleanliness. Initially the bead is spherical, but it becomes hemispherical after melting and alloying. Great care has to be taken in the control of temperature and time for the alloying process. Too short a time or too low a temperature will preclude the formation of a good low-resistance contact. Too long an alloying time at too high a temperature causes too much dissolution of the metals in the GaAs with the formation of a temperature-dependent contact resistance (Bolton and Jones 1969; Harris, Nannichi, Pearson, and Day 1969), which is presumably caused by crystal dislocations, and the introduction of undesired electron traps in the forbidden energy gap of the semiconductor. Contact to the alloyed bead may be made with a thermocompression-bonded gold wire or simply by pressure.

As a device develops towards maturity better techniques are used. The contacts are formed by photolithographic definition of their position and evaporation of suitable metals, which are subsequently alloyed as was done for the bead contact (Fig. 6.3 (b)). Suitable metals are 50 nm of Sn followed by 500 nm of Ag, which prevents the Sn forming into droplets during the alloying cycle. Another popular contact alloy is AuGe with a small amount of Ni. The Ni is evaporated after the Au and Ge, and it is necessary to avoid droplet formation during alloying. The individual devices are obtained by cleaving the epitaxial layer after contact formation, so that the labour and the cost are reduced compared with the individually-contacted bead devices. Contact is made to the alloyed metals with a thermocompression-bonded gold tape or wire with a diameter of $\sim 50\ \mu$m.

It is very difficult to avoid degradation of the alloyed contact referred to earlier, but considerable progress has been made recently (Basterfield, Board, and Josh 1972). With the penalty of somewhat greater cost many manufacturers prefer to grow an n^+ surface epitaxial layer so that an alloyed-metal contact can be made outside the active region and contact effects are avoided. Such a device is illustrated in Fig. 6.3 (c), and it includes an n^+ buffer layer between the substrate and active region.

The mesa device profiles in Figs. 6.3 (b) and 6.3 (c) are defined by the contacts and suitable etchants and avoid r.f. power loss in the adjacent inactive regions of the device illustrated in Fig. 6.3 (a). Mesa profiles are a practical necessity for the inverted devices described later.

6.3. Thermal design
In addition to the good electrical contacts a device must have a good thermal contact with its environment to avoid it's destruction by overheating. For most geometries the detailed solution of the heat-flow problem through a small active device into a massive heat-sink is involved and complicated. However, useful results may be obtained with the aid of simple approximations, and such an approach will be taken here. The removal of heat will be considered to occur from one end of the device only. Brief comments on double-sided heat-sinking will follow. Some differences will arise in the detailed form of the expressions when compared with those given elsewhere, but similar numerical values will result.

Pulsed devices whose length is limited either by transit-time effects or material homogeneity have as large an area as possible in order to generate high power. A limit is reached with a low-field device resistance of approximately 1 Ω (see §7.3) to avoid considerable power loss in matching to the electromagnetic circuit. For a given device length and resistance it is advantageous to have as large a device area as possible by increasing the resistivity so that this area may make contact with the heat-sink and more efficiently allow heat flow. A lower resistivity limit is set by efficiency considerations when the doping–length product $N_D l$ falls towards $10^{16} \, \text{m}^{-2}$ owing to reduction of the depth of r.f. current modulation in the device. A lower limit of $N_D l \simeq 10^{17} \, \text{m}^{-2}$ must usually be set when the ultimate efficiencies are required in high-peak-power pulsed devices operating at a low duty cycle. $N_D l \simeq 10^{16} \, \text{m}^{-2}$ is often a better compromise in c.w. devices where high-power output is of more importance than efficiency. These length (or frequency) and $N_D l$ requirements place limitations on the device radius r, as shown in Fig. 6.4. For comparison purposes, a doping density of $10^{21} \, \text{m}^{-3}$ corresponds approximately to a low-field resistivity of $0.01 \, \Omega \text{m}$ in good-quality GaAs. To a good approximation the aspect ratio $2r/l$ of the device is much greater than unity, so that it is essentially a thin sheet in intimate contact with the heat-sink.

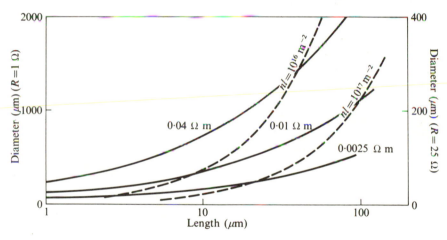

Fig. 6.4. Relationship between length and diameter to produce a device with a low-field resistance of $1 \, \Omega$ or $25 \, \Omega$ with the additional constraint of constant resistivity or constant $N_D l$ product.

If the device has been well fabricated, so that interfacial and contact electrical resistance is negligible, all the heat is dissipated in the active region of the device. The composite heat-flow geometry may be approximated by the series connection of a one-dimensional active device region and a heat-sink with spherically-symmetrical heat flow, as illustrated in Fig. 6.5. If W is the total power dissipation in the device, the temperature at the interface (the 'zero-level' for the active region) is the same as the temperature at a distance $r/\sqrt{2}$ from a point source on the surface of a semi-infinite heat-sink when the thermal power flow from this source is W. Ambient temperature occurs at infinity. The factor $1/\sqrt{2}$ is necessary to make the surface contact areas of the disc and hemisphere equal.

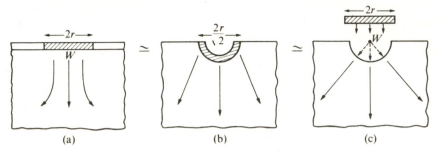

Fig. 6.5. Illustration of the approximation used in the device geometry to simplify calculations of thermal behaviour.

6.3.1. Static temperature distribution and the thermal resistance

The solution to both components of the heat-flow problem is well known (Carslaw and Jaeger 1959) and involves solving, respectively, the heat-flow equations in one dimension with uniform heating and in three dimensions with spherically symmetric heat flow from a point source. If W is dissipated uniformly over the device of length l and radius r, the temperature θ_x in the active region at a distance x from the interface at temperature θ_s is

$$\theta_x - \theta_s = \frac{Wx(2l - x)}{2\pi r^2 l K_a}, \tag{6.1}$$

where K_a is the active-region thermal conductivity.

The maximum active-region temperature is θ_1 given by

$$\theta_1 - \theta_s = \frac{Wl}{2\pi r^2 K_a}. \tag{6.2}$$

Heat flow through the substrate to an ambient temperature θ_A at infinity yields

$$\theta_s - \theta_A = \frac{W}{\sqrt{2}\pi r K_s}, \tag{6.3}$$

where K_s is the substrate thermal conductivity.

We may now use eqns (6.2) and (6.3) to illustrate the thermal design of the device. It is advantageous to make r as large as possible subject to the $N_D l$ limitations illustrated in Fig. 6.4. If $K_s = K_a$,

$$\frac{\theta_s - \theta_A}{\theta_1 - \theta_s} = \sqrt{2}\frac{r}{l}. \tag{6.4}$$

Usually $r \gg l$ so that most of the temperature difference occurs across the substrate. $K_s = K_a \simeq 50 \text{ W m}^{-1}\text{K}^{-1}$ corresponds to a GaAs device whose GaAs

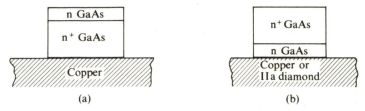

Fig. 6.6. In (a) heat flow is through the substrate with high thermal resistance. The flip-chip mounting of (b) allows heat flow directly into the heat-sink.

substrate is bonded to a distant heat-sink (Fig. 6.6 (a)). For a typical low-power X-band c.w. device with $l = 10 \ \mu$m, $r = 50 \ \mu$m, and $W = 1$ W,

$$\theta_1 - \theta_s = 13 \text{ K}$$

and

$$\theta_s - \theta_A = 90 \text{ K} ,$$

yielding a thermal resistance of 103 K W^{-1}. The temperature-dependence of the GaAs thermal conductivity has been neglected (Bulman *et al.* 1972), and a somewhat higher ($\sim$25 percent) thermal resistance would occur in practice.

The structure illustrated in Fig. 6.6 (a) cannot dissipate more than 1 or 2 W, but considerable improvement follows replacement of the substrate by a material with much higher thermal conductivity. If $K_s = 385$ W m^{-1} K^{-1}, corresponding to copper, we have

$$\frac{\theta_s - \theta_A}{\theta_1 - \theta_s} = \sqrt{2} \frac{K_a}{K_s} \cdot \frac{r}{l} = 0 \cdot 92$$

for the above dimensions. There is now roughly equal temperature difference across each part of the device, and the thermal resistance has been reduced to approximately 25 K W^{-1}. Increase of the device radius (without reduction of device doping) to 100 μm will reduce the thermal resistance to approximately 9 K W^{-1} and the maximum power dissipation is 10 W to 15 W. The practical realization of this structure is the 'flip-chip' or 'integral heat-sink' device illustrated in Fig. 6.6 (b). Usually a thin film (a few micrometres) of gold is electroplated onto the n-type GaAs, which is then ultrasonically or thermocompression bonded to the metal heat-sink, which has also been electroplated with gold.

Further improvement may be obtained for exotic applications by replacing the copper with type-IIa diamond for which $K_s = 2000$ W m^{-1} K^{-1} at 300 K. A dissipation of approximately 30 W may be achieved in practice. Thermal resistance for various geometries of GaAs device on the three types of heat-sink are summarized in Table 6.1.

TABLE 6.1

Substrate	GaAs	R_T (KW^{-1}) Cu	IIa diamond
$l = 10 \ \mu m$ $r = 50 \ \mu m$	103	25	15
$l = 100 \ \mu m$ $r = 50 \ \mu m$	217	139	130
$l = 100 \ \mu m$ $r = 500 \ \mu m$	92	13	3·5

The thermal resistances in the third row of Table 6.1 look attractively small, but they are counteracted by higher power dissipation to maintain the same bias field. For example, the devices in the third row would have 1000 times the power dissipation of the device in the first row if they had equal doping densities or 100 times if they had equal $N_D l$ products.

6.3.2. Transient and a.c. thermal response

Time-varying thermal behaviour is important both in pulsed devices and in modulated c.w. devices. Following a step-function change of thermal input power ΔW the temperature of any region will change at a rate dictated by its thermal capacity and the net flow of thermal energy into it. As the interface temperature between the device and substrate changes, a constant-temperature surface will 'diffuse' outwards from the interface, but the rate of temperature change will be smaller owing to the larger thermal capacity of the heat-sink.

The transient response of the one-dimensional active region is described through Fourier analysis of its temperature distribution (Carslaw and Jaeger 1959). Each Fourier component has a characteristic exponential relaxation time to the new steady state. As may be expected, the component with the longest 'wavelength' has the longest relaxation time. The temperature distribution in the active region may be described within an accuracy of 10 percent by this one component, which has a time variation given by

$$\theta_x - \theta_s = \frac{\Delta W l}{2\pi r^2 K} \left\{ \frac{x}{l^2}(2l - x) - \frac{32}{\pi^3} \cdot \sin\left(\frac{\pi x}{2l}\right) \exp\left(\frac{-\kappa \pi^2 t}{4l^2}\right) \right\}, \quad (6.5)$$

i.e.

$$\theta_1 - \theta_s = \frac{\Delta W l}{2\pi r^2 K} \left\{ 1 - \frac{32}{\pi^3} \exp\left(\frac{-\kappa \pi^2 t}{4l^2}\right) \right\}. \quad (6.6)$$

$\kappa = K/\rho c$ is the thermal diffusivity, ρ is the density, and c is the specific heat. We will see shortly that the substrate response is slower than the active-region response, so that for high-peak bias power applied for short duration we can take $\theta_s \simeq$ constant. If $\Delta W l/2\pi r^2 K$ is much larger than the allowable temperature rise the pulse must be applied for less than the exponential relaxation time $4l^2/\kappa \pi^2$. The temperature will be linearly related to time by (neglecting the difference between $32/\pi^3$ and unity)

$$\theta_1 - \theta_s = \frac{\Delta W}{\rho c} \cdot \frac{\pi^2}{8} \cdot \frac{t}{\pi r^2 l}.$$

$\rho c \pi r^2 l$ is the thermal capacity S so

$$\theta_1 - \theta_s = \frac{\Delta W}{S} \cdot \frac{\pi^2}{8} \cdot t. \tag{6.7}$$

A typical pulsed transit-time device operating in X-band will have $\Delta W \simeq$ 100 W, a diameter of 300 μm, and a length of 10 μm, so that $S \simeq 10^{-6}$ JK^{-1}. Under these conditions $\theta_1 - \theta_s \simeq 100$ K μs^{-1}, so a pulse length less than approximately 1 μs is required. The power dissipation during the pulse is approximately 0·1 MW mm^{-3}.

Calculations are more difficult when the bias power input is not great enough to make the ultimate temperature change, predicted by eqn (6.6) alone, exceed the maximum allowable device temperature. Account must be taken of the rise of temperature at the device—substrate interface in order to estimate the maximum allowable pulse duration. As the temperature change diffuses through the substrate it encompasses an ever increasing bulk of material. The interface temperature changes according to (Carslaw and Jaeger 1959)

$$\theta_s = \frac{\Delta W}{\sqrt{2}\pi K r} \operatorname{erfc} \left(\frac{r^2}{8\kappa t} \right)^{\frac{1}{2}} \text{ K}. \tag{6.8}$$

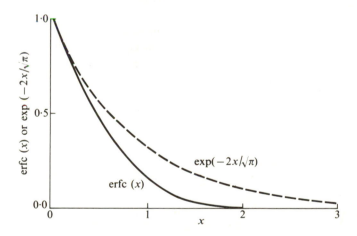

Fig. 6.7. Comparison of the error function erfc(x) and an exponential to which it approximates for small x.

This function is illustrated in Fig. 6.7. It has an initially rapid variation, followed by a slow approach to the steady state, and is 'squarer' than an exponential function. θ_s completes half its over-all change in a time of approximately $r^2/2\kappa$. The ratio of this 'half-life' to the time constant of eqn (6.6) is $8l^2\kappa_s/\pi^2 r^2\kappa_a$, where κ_s is the diffusivity of the substrate and κ_a is the diffusivity of the active device. Usually this ratio is much less than unity, so the essentially instantaneous and complete change of temperature in the active region must be added to the time-varying temperature of the interface. The pulse length is limited by the time at which the highest instantaneous temperature reaches its maximum allowable value. The 'scale factor' of the pulse length is the half-life $r^2/2\kappa$, and representative values are shown in Table 6.2.

TABLE 6.2

$r^2/2\kappa$ (μs)

$r(\mu m)$	GaAs	Cu	IIa diamond
50	40	11	1·1
250	1000	275	27·5

As a simple design example, low duty-cycle operation for devices with a much larger ultimate temperature difference across the substrate than across the active device could have a peak bias-power input equal to twice the c.w. power input for the times in Table 6.2.

We have not dealt with the cooling period between pulses. In general, this problem of the temporal decay of an arbitrary temperature distribution is an involved problem, whose solution is described by Carslaw and Jaeger (1959). In practice, the shortest time between pulses is determined empirically, but it is possible to make a few simple observations. The mean power input must always be less than is achievable under c.w. conditions because the peak device-temperature is essentially equal to the maximum continuous temperature. However, the mean dissipated power of pulsed operation will approach that of c.w. operation if the pulse-width is short enough ($\sim$ 100 ns typically) to avoid a large temperature excursion. The cooling time between pulses is of the order of $r^2/2\kappa$ (De Sa and Hobson 1971) because the heat has to be removed through the substrate. Under long-pulse conditions ($\simeq r^2/2\kappa$) the calculations are more involved, but the cooling time will be longer than the heating time when the peak bias power is greater than twice the c.w. maximum power, because the asymptotic temperature change of the latter is greater than that of the former. At this point empirical techniques become invaluable.

In principle, double-sided heat-sinking techniques should reduce the thermal resistance and transient-temperature excursions (for substrate-limited pulse lengths) by a factor of 2. Great difficulty is encountered in constructing a second continuous conducting path which will not crush the device, but the relatively-small gold wire which makes contact to the free device surface can have sufficient thermal capacity to give a useful increase of pulse length for substrate-limited pulse lengths.

When modulated c.w. oscillators are under consideration it is often more convenient to express the above results in the frequency domain instead of the time domain. These considerations apply not only to bias-circuit modulated oscillators but also to other forms of modulation which cause inadvertent modulation of the oscillator loading. The bias current has some dependence on this loading, so a sympathetic bias-power dissipation may be set up.

If the a.c. bias-power input is $\Delta W_0 \cos \omega t$ (Carslaw and Jaeger 1959),

$$\theta_1 - \theta_s = \frac{\Delta W_0 l}{2\pi r^2 K} \frac{1}{\sqrt{(\omega^2 \tau^2 + 1)}} \cos \{\omega t - \tan^{-1}(\omega \tau)\}, \qquad (6.9)$$

where $\tau = 4l^2/\kappa \pi^2$, and

$$\theta_s = \frac{\Delta W_0}{\sqrt{2\pi Kr}} \exp\left(\frac{-\omega r^2}{4\kappa}\right)^{\frac{1}{2}} \cos\left\{\omega t - \left(\frac{\omega r^2}{4\kappa}\right)^{\frac{1}{2}}\right\}. \qquad (6.10)$$

Eqn (6.10) is only valid for $\omega \ll 1/\tau$.

6.4. Packaging
The bare GaAs device is easily damaged mechanically or by impurities carried in the atmosphere, so it is necessary to provide it with a robust, hermetically sealed package so that it may be easily handled and incorporated in a variety of microwave circuits. Inevitably the electrical performance of the device is restricted (Taylor, Fray, and Gibbs 1970) by the addition of spurious circuit elements such as the stray parallel capacity and series inductance of the connecting gold wires (Owens and Cawsey 1970). The package size must be reduced with increasing frequency as these effects become increasingly serious. The smaller packages are most expensive owing to the intricate handling that is required, and they place some limitations on the maximum power-handling capacity by restricting the heat-sink dimensions in the vicinity of the diode and by introducing spurious interfacial thermal resistance. The cross-section of some packages are illustrated in Fig. 6.8. In some applications the package may be too restrictive electrically and the device will have to be mounted directly in the circuit with loss of flexibility in handling.

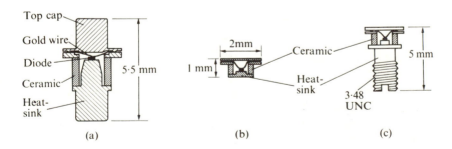

Fig. 6.8. Three commonly used packages with hermetic seals: (a) S4; (b) S1; (c) Interceram AV170.

6.5. Device reliability

When properly constructed, mounted, and operated, c.w. transferred-electron devices have lives in excess of 30 000 hours. This estimate is largely set by the available time since good-quality devices became available, and problems with failures which can be attributed to power-supply failure. Devices with a large contact resistance often have shorter lives, measured in thousands of hours or less.

A common cause of premature failure in c.w. devices is overheating, followed by melting of the gold contact wire so that the device becomes open-circuited. The time required for this to occur is somewhat less than a millisecond (see the earlier thermal analysis), so it is important that power supplies do not have a voltage overshoot when switched on. Another form of overheating failure occurs, apparently inexplicably, after a device has been operating for a period often of several hours. It is usually traceable to an imperfect thermal contact of the device package with the large surface area of the microwave circuit which transfers heat to the atmosphere. The device package must be firmly pressed into its heat-sink and some silicone grease used to improve the thermal contact. For continuous dissipation in excess of 1 W it is necessary to positively clamp the package to the heat-sink using a tapered mount such as illustrated in Fig. 6.9.

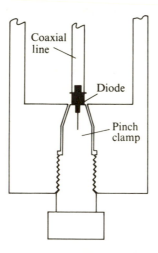

Fig. 6.9. Pinch clamp necessary for power dissipation in excess of 1 W.

Short-circuit diode failure can occur under pulsed conditions if the metal contact on the epitaxial layer is operated as the anode. Positively-charge metal ions from the contact drift through the diode under the influence of the bias electric field and form a short-circuiting 'spike' through the device. Under c.w. conditions the process may take less than 1 s, but the high current and excessive heating that follows breakdown often causes melting of the gold wire and open-circuit failure. When the metal—epitaxial contact is correctly operated as the cathode, similar conducting channels can result from material damage following avalanche breakdown when the pulsed bias voltage is too large.

7 **Circuit properties**

7.1. Introduction

The majority of circuit design for microwave semiconductor oscillators and amplifiers is an empirical affair which is guided by a few fundamental principles (Montgomery *et al.* 1948) and a lot of painful experience. Only the former are summarized here, but further practical details have been collected by Bulman *et al.* (1972). Initially, a lumped equivalent circuit is required to avoid the necessity of solving Maxwell's equations for all problems.

7.2. The parallel (or series) equivalent circuit

In developing an equivalent circuit for a diode and cavity a prime consideration is simplicity of interpretation, particularly with respect to change of frequency. The dimensions of a practical diode are much smaller than the wavelength of any signals to which it is coupled, so a lumped-circuit model is appropriate. The parallel flow of current through the conductive loss and capacity of a uniform dielectric material when both are naturally subject to the same electric fields suggests a parallel equivalent circuit. The conductance becomes negative and the electric field is non-uniform in transferred-electron devices, so that the choice is not so clear. When operating in their optimum frequency range many devices have a negative conductance G_D which is appreciably smaller than the parallel susceptance ωC_D and is substantially independent of frequency. A parallel nature arises from the terminal requirement that the waveform of the r.f. voltage is the same for real and imaginary components of the r.f. current. The admittance Y_D and the equivalent impedance X_D are given by

$$Y_D = G_D + j\omega C_D,$$

$$X_D \simeq \frac{G_D}{\omega^2 C_D{}^2} + \frac{1}{j\omega C_D}.$$

The frequency-dependence of the real part of X_D is clearly more complicated than the of Y_D, so the parallel equivalent is often chosen. The choice is not always clearcut, and it may be more convenient to use the series representation when, for example, the diode is mounted in series with a large inductance.

The cavity circuit is often a 'jungle of brasswork' which we know is connected to the diode at two terminals. There are circuit theorems to show that any lossless circuit may be represented by a series connection of parallel LC circuits or a parallel connection of series LC circuits (Montgomery *et al.* 1948). Each LC pair is required to describe a pole and zero (including those at zero and infinite frequency) of the reactance frequency response. If the cavity has been carefully designed the resonant frequencies will be widely spaced and no more than one will exist within the frequency range of interest. In such a delightfully simple situation the description may be sufficiently accurate with one parallel or one series equivalent LC circuit. Analysis is simplified if both the diode and its controlling circuit have the same form. The parallel equivalent illustrated in Fig. 7.1 will generally be used here. Power is coupled to the external circuit by controlled leakage of the r.f. electric field and/or the magnetic field. This useful loss is represented by the conductance G_L connected to the circuit through a transformer. Alternatively, it may be transformed to the diode terminals as G_L and connected as shown. G_W is the spurious-loss conductance of the cavity walls.

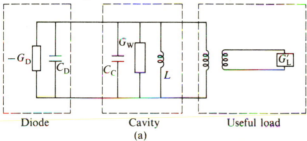

Diode   Cavity   Useful load

(a)

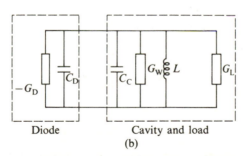

Diode   Cavity and load

(b)

Fig. 7.1. Equivalent circuit of device, controlling circuit, and load.

7.3. The effect of circuit losses on the maximum r.f. power generation

In discussing the Pf^2 limitations it was pointed out that excessive power loss occurs in the cavity walls if the device impedance is low. With all other parameters unaltered, an increase of diode area (and admittance) will give a proportional increase in the maximum r.f. power-generating capability. The increased admit-

Circuit properties

tance requires the device to 'sit' in a lower-impedance position of the standing-wave pattern which exists in the cavity. The diode length is unaltered so the terminal r.f. voltage is independent of area. Consequently, the maximum voltage and current amplitudes of the standing-wave pattern increase and do so in proportion to the admittance when it is much greater than the characteristic admittance of the cavity. The r.f. power lost in the cavity walls is proportional to the square of the maximum r.f. voltage or current amplitude. Therefore the ratio of the r.f. power lost in the walls to that generated by the device increases with increasing device admittance. When the admittance is large enough a condition will be reached in which a further increase of diode area will cause an increase of the total r.f. power generated by the device, but the net r.f. power usefully coupled out of the cavity will decrease.

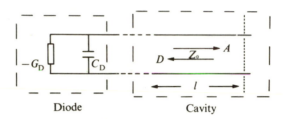

Fig. 7.2. A transmission-line equivalent of the controlling circuit as used to calculate power loss in the cavity walls.

These effects may be illustrated and roughly estimated by a simplified model in which a diode of admittance $-G_D + j\omega C_D$ is connected across one end of a transmission line of characteristic impedance Z_0 (Fig. 7.2). Coupling of useful power P_{out} to the external circuit is carried out at the other end of the transmission line with an appropriate reflective tuner a distance l from the diode. The transmission-line circuit of Fig. 7.2 is equivalent to the lumped equivalent circuit of the cavity in Fig. 7.1. The diode has a Q-factor Q_D defined as $\omega C_D / G_D$ and is placed at $x = 0$ in the transmission line, whose voltages and currents are given by (Montgomery *et al.* 1948)

$$V = \exp(j\omega t)\{A\exp|-(\alpha + jk)x| + D\exp(j\theta)\exp|(\alpha + jk)x|\}, \quad (7.1)$$

$$Z_0 i = \exp(j\omega t)\{A\exp|-(\alpha + jk)x| - D\exp(j\theta)\exp|(\alpha + jk)x|\}. \quad (7.2)$$

A and D are real voltage amplitudes, α is the voltage attenuation coefficient caused by the transmission-line wall losses, k is the wave number, and θ is a convenient phase factor, determined by the admittance of the diode, which relates the phases of the incoming and outgoing waves at $x = 0$. θ will be determined by making the transmission line a conjugate match to the diode admittance as required for resonance. In practice, both the position of the tuner and the magnitude of its reflection would have to be adjusted in order to obtain optimum power output at the desired wavelength. The input admittance of the transmission line is obtained from the ratio of eqns (7.1) and (7.2) and the conjugate match criterion at $x = 0$ gives

$$Q_D = \frac{2AD\sin\theta}{(A^2 - D^2)}.$$ (7.3)

Also

$$|V_0|^2 = A^2 + 2AD\cos\theta + D^2$$ (7.4)

and

$$P_G = \frac{|V_0|^2 G_D}{2} = \left(\frac{A^2 - D^2}{2Z_0}\right),$$ (7.5)

where $|V_0|$ is the r.f. voltage amplitude at the diode terminals when adjusted for optimum r.f. power generation P_G. It is assumed that the output tuner is always adjusted for optimum power generation so that

$$P_{out} = P_G - P_W,$$ (7.6)

where P_W is the r.f. power lost in the cavity walls. The proportion of r.f. power lost in travelling a distance l in the lossy line is $\{1 - \exp(-2\alpha l)\}$, so that the total wall loss is

$$P_W = \frac{A^2}{2Z_0}\{1 - \exp(-2\alpha l)\} + \frac{D^2}{2Z_0}\{\exp(2\alpha l) - 1\}.$$

For the generally applicable situation that $\alpha l \ll 1$, the above equation simplifies to

$$P_W = (A^2 + D^2)\alpha l/Z_0.$$ (7.7)

As the device admittance is increased, the standing-wave ratio in the cavity tends to infinity and $A/D \to 1$. The ratio of P_W/P_G, taken from eqns (7.5) and (7.7), increases as A/D increases towards unity. This is the essential matching limitation referred to at the beginning of this section.

By eliminating θ from eqns (7.3) and (7.4) and A and D with eqns (7.5) and (7.7) we have

$$2P_G Z_0{}^2 G_D(1 + Q_D{}^2) = \left(\frac{2Z_0 P_W}{\alpha l} - |V_0|^2\right).$$ (7.8)

Eliminate P_W from eqns (7.6) and (7.8):

$$P_{out} = P_G\{1 - \alpha l Z_0 G_D(1 + Q_D{}^2)\} - \frac{|V_0|^2 \alpha l}{2Z_0}.$$

For all high-power cases of interest $G_D \gg \alpha l/Z_0$ (see below), therefore

$$P_G \gg \frac{|V_0|^2 \alpha l}{2Z_0},$$

Circuit properties

and

$$P_{\text{out}} \simeq P_G \{1 - \alpha l Z_0 G_D (1 + Q_D{}^2)\}. \tag{7.9}$$

Eqn (7.9) shows that $P_{\text{out}} \simeq P_G$ for small-area diodes (G_D small) but $P_{\text{out}} = 0$ when the diode area is large enough to give

$$G_D = \frac{1}{\alpha l Z_0 (1 + Q_D{}^2)}.$$

In this condition the cavity-wall loss alone is the optimum load. The maximum power output is obtained when half the total power generated is dissipated in the cavity walls. This optimum output is

$$(P_{\text{out}})_{\text{opt}} = \frac{|V_0|^2}{8 \alpha l Z_0 (1 + Q_D{}^2)},$$

and is obtained when

$$(G_D)_{\text{opt}} = \frac{1}{2 \alpha l Z_0 (1 + Q_D{}^2)}.$$

For a Gunn diode operating in X-band the following numerical values would be appropriate if it were mounted directly across a full-height waveguide with a bias voltage $\simeq 15$ V: $Q_D \simeq 3$; $Z_0 \simeq 500$ Ω; $\alpha l \simeq 10^{-3}$; $|V_0| \simeq 10$ V; therefore,

and
$$\left.\begin{array}{l} (P_{\text{out}})_{\text{opt}} = 2 \cdot 5 \text{ W} \\ (G_D)_{\text{opt}} = 10^{-1}\text{S} \end{array}\right\} \text{waveguide}$$

and half the total generated power would be dissipated in the cavity walls. Corresponding quantities for a coaxial transmission line would be $Z_0 \simeq 50$ Ω; $\alpha l \simeq 20 \times 10^{-3}$; therefore,

and
$$\begin{array}{l} (P_{\text{out}})_{\text{opt}} = 1 \cdot 25 \text{ W} \\ (G_D)_{\text{opt}} = 5 \times 10^{-2}\text{S} \end{array} \text{ coaxial}.$$

In both of the above cavity situations $(G_D)_{\text{opt}}$ is the modulus of the negative conductance of the device when generating maximum power. The optimum value is typically 20 to 40 times smaller than the low-field positive conductance (Copeland 1967a, b; Freeman and Hobson 1973), so that the low-field resistance of high-power devices is often approximately 1 Ω.

In practice, it is usually impossible to mount the diode directly at the end of a transmission line, and account must be taken of mounting parasitics. Detailed calculations will reveal different numerical results even though we may expect the same order of magnitude. $|V_0|$ may be 3 or 4 times larger than assumed above for pulsed transit-time devices, so we may expect peak power outputs up to 25 W from a single diode, in good agreement with experiment.

Transit-time devices designed for higher frequencies have a smaller $(P_{\text{out}})_{\text{opt}}$

owing to the direct dependence of V_0 on the device length, as discussed in § 4.5 on the Pf^2 limitations. There will be a more drastic reduction of $(P_{out})_{opt}$ and $(G_D)_{opt}$ if Q_D is higher for higher-frequency devices. However, good design of the circuit to avoid spurious capacity in parallel with the device will maintain Q_D approximately frequency-independent, as may be seen from the following argument. If we assume that the optimum negative conductance of an oscillator is 20 times smaller than the low-field conductance (Copeland 1967a, b; Freeman and Hobson 1973) and the device capacity is approximately equal to the low-field capacity, we have

$$Q_D = 40\pi f \epsilon \epsilon_0 / n e \mu \,,$$

where μ is the low-field mobility. If we take $\mu \simeq 0.7 \text{ m}^2\text{V}^{-1}\text{s}^{-1}$ and $\epsilon = 12.5$,

$$Q_D = \frac{1.24 \times 10^{11}}{n/f} \,,$$

where n/f is in units of s m^{-3}.

For a sinusoidal LSA device $n/f \simeq 5 \times 10^{10} \text{s m}^{-3}$. C.w. transit-time devices are usually designed with an $N_D l$ product $\simeq 10^{16} \text{m}^{-2}$ (see § 6.3), and they operate at a frequency given by

$$f \simeq v/l \,,$$

where $v \simeq 10^5 \text{ m s}^{-1}$ and l is the device length. Therefore the n/f ratio of a c.w. transit-time device is $\simeq 10^{11} \text{s m}^{-3}$. It can be seen that both types of device have a $Q_D \simeq 1$, but the capacity is usually somewhat greater than the low-field capacity (Copeland 1967a, b; Hobson 1967) and typically $2 < Q_D < 5$. The Q_D of LSA relaxation oscillators operating with $n/f \simeq 2 \times 10^{11} \text{s m}^{-3}$ will be a little smaller.

7.4. Circuit stabilization of frequency or amplitude fluctuations
Referring to Fig. 7.1 the steady-state angular frequency ω and power output of an oscillator are determined by the conditions

$$B = B_C(\omega) + B_D(\omega, V_0) = 0 \,, \tag{7.10}$$

$$G = G_L(\omega) + G_W(\omega) + G_D(\omega, V_0) = 0, \tag{7.11}$$

where B_C and B_D are the susceptances and G_L, G_W, and G_D are the conductances of the circuit, wall loss, and diode, respectively. Their functional dependence on ω and the r.f. voltage amplitude V_0 are indicated. If small changes of susceptance and conductance, ΔB_S and ΔG_S respectively, are introduced to the circuit either desirably or undesirably there will be a change of V_0 and ω in order to maintain the equalities of eqns (7.10) and (7.11). If ΔB_S or ΔG_S are device instabilities the circuit must be designed to make ΔV_0 and $\Delta\omega$ as small as possible. When ΔB_S or ΔG_S are controlled, to cause modulation, it is desirable to make ΔV_0

and $\Delta\omega$ as large as possible. The circuit-design criteria are derived from the changes ΔB and ΔG in eqns (7.10) and (7.11). We have

$$\Delta B_S + \Delta B = 0 \, ,$$

$$\Delta G_S + \Delta G = 0 \, ,$$

i.e.

$$\Delta B_S = -\left(\frac{\partial B}{\partial \omega}\right)\Delta\omega - \left(\frac{\partial B}{\partial V_0}\right)\Delta V_0 \, , \tag{7.12}$$

$$\Delta G_S = -\left(\frac{\partial G}{\partial \omega}\right)\Delta\omega - \left(\frac{\partial G}{\partial V_0}\right)\Delta V_0 \, . \tag{7.13}$$

Eqns (7.12) and (7.13) may be rearranged to give

$$\Delta\omega = \frac{-\Delta B_S \left/\left(\frac{\partial B}{\partial V_0}\right)\right. + \Delta G_S \left/\left(\frac{\partial G}{\partial V_0}\right)\right.}{\left(\frac{\partial B}{\partial \omega}\right)\left/\left(\frac{\partial B}{\partial V_0}\right)\right. - \left(\frac{\partial G}{\partial \omega}\right)\left/\left(\frac{\partial G}{\partial V_0}\right)\right.} \tag{7.14}$$

and

$$\Delta V_0 = \frac{-\Delta B_S \left/\left(\frac{\partial B}{\partial \omega}\right)\right. + \Delta G_S \left/\left(\frac{\partial G}{\partial \omega}\right)\right.}{\left(\frac{\partial B}{\partial V_0}\right)\left/\left(\frac{\partial B}{\partial \omega}\right)\right. - \left(\frac{\partial G}{\partial V_0}\right)\left/\left(\frac{\partial G}{\partial \omega}\right)\right.} \tag{7.15}$$

Usually $\partial G/\partial\omega$ is small compared with $\partial B/\partial\omega$ but $\partial B/\partial V_0$ and $\partial G/\partial V_0$ are comparable (Bestwick, Drinan, Hobson, Robson, Thomas, and Tozer 1973), so that eqns (7.14) and (7.15) simplify to

$$\Delta\omega = \left\{-\Delta B_S + \frac{(\partial B/\partial V_0)}{(\partial G/\partial V_0)}\Delta G_S\right\}\left/\left(\frac{\partial B}{\partial \omega}\right)\right. , \tag{7.16}$$

$$\Delta V_0 = \left\{-\Delta G_S + \frac{(\partial G/\partial \omega)}{(\partial B/\partial \omega)}\Delta B_S\right\}\left/\left(\frac{\partial G}{\partial V_0}\right)\right. \simeq -\Delta G_S\left/\left(\frac{\partial G}{\partial V_0}\right)\right. \tag{7.17}$$

The fluctuation of frequency will be small if $\partial B/\partial\omega$ is large. This condition is a circuit requirement for large stored energy (Montgomery *et al.* 1948). Large $\partial B/\partial\omega$ will also reduce amplitude fluctuations caused by susceptance fluctuations, but the overriding requirement is for a large $\partial G/\partial V_0$ a quantity dependent only on the diode and its working conditions.

Returning to $\partial B/\partial\omega$, the r.f. energy stored in the cavity is E_0, given by

$$E_0 = \frac{V_0{}^2}{4}\frac{\partial B}{\partial \omega} \, , \tag{7.18}$$

as may be verified by reference to the parallel LC circuit or, more rigorously, in

Montgomery *et al.* (1948). In the parallel *LC* circuit, $\partial B/\partial \omega$ is equal to $2C$, so high stability requires large parallel capacity and small parallel inductance. Large $\partial B/\partial \omega$ and E_0 place a requirement for a large unloaded-cavity Q-factor Q_0.

$$Q_0 = \frac{2\pi \text{ (r.f. energy stored in cavity)}}{\text{(total energy dissipation per cycle)}}$$

$$= \left\{ 2\pi \cdot \frac{V_0{}^2}{4} \cdot \frac{\partial B}{\partial \omega} \right\} \bigg/ \left\{ \frac{1}{f} \cdot \frac{V_0{}^2}{2} \cdot G_W \right\} = \frac{\omega}{2G_W} \cdot \frac{\partial B}{\partial \omega}, \qquad (7.19)$$

where G_W is the appropriate loss conductance. If the design requirement is for easy frequency variation the stored energy (and Q_0) must be as small as possible.

7.5. Q-factor in non-linear circuits

The Q-factor is a convenient quantity in linear-circuit analysis for the description of circuit bandwidth, oscillatory-decay time constants, and proportional power losses in filters, etc. Errors arise in the application of the concept to oscillators which, of necessity, are large-signal, non-linear devices. Before progressing further it is necessary to identify the external, the loaded, and the unloaded Q-factors. They are each associated with the same stored energy but their loss conductances are, respectively, the useful external load, the wall losses plus this external load, and the wall losses alone. Occasionally, the negative Q is used to define the relationship of the small-signal negative conductance and stored energy for use in the initial growth of oscillation in pulsed oscillators, but it will not concern us here.

The Q-factor of a given device in a given circuit is a function of the voltage amplitude. The most common mistake, in much published work, is to relate the cavity conditions of an oscillator to its stability by the loaded or external Q-factor alone. The definition of Q (in the previous section) clearly includes both stored r.f. energy and dissipated energy, and is ambiguous. The important circuit consideration for high stability (defined as the inverse of the frequency fluctuation) was shown in the previous section to be its proportionality to large $\partial B/\partial \omega$, or stored energy, alone. It is possible to vary Q without changing $\partial B/\partial \omega$ when any modification of stability will be caused by alteration of $\partial G/\partial V_0$ or of the magnitude of the prime fluctuation. Stability will not be proportional to Q in general. A complete description of oscillator-cavity conditions can only be given by specifying the stored energy and the load conditions. Practical necessity may cause these two quantities to be replaced by the mixed terms of loaded Q-factor and the undercoupled or overcoupled power output relative to the maximum possible power.

7.6. Frequency locking and frequency pulling

7.6.1. Introduction to the effects

A small quantity of incoming microwave power may have a significant effect on

Circuit properties

the frequency and the much larger power leaving an oscillator. Two processes, frequency locking and frequency pulling, result from this interaction. Synchronization or frequency locking will occur if the incoming power is injected by a separate oscillator within a finite frequency range overlapping that of the free-running oscillator. A similar process derives the incoming power from a reflection, spurious or otherwise, in the output circuitry, and the frequency of the free-running oscillator is pulled according to the phase and amplitude of the reflected signal. As the reflector is moved along the output transmission line the frequency excursion repeats every half wavelength, as illustrated in Fig. 7.3.

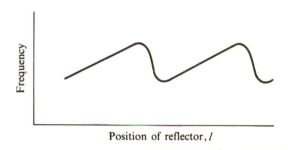

Fig. 7.3. Frequency pulling by an external, reflecting discontinuity.

Frequency or phase control may be achieved with frequency locking, whereas frequency pulling is of interest to determine the maximum mismatch which may be tolerated by an oscillator with a given frequency specification. Both are commonly used to determine the external Q-factor of an oscillator. When frequency locking has been achieved it is essentially the same process as frequency pulling. Only one frequency exists in the system but there are slight differences in the transmission paths of the external power. The input power P_{in} and output power P_{out} are often separated by a circulator for frequency locking (Fig. 7.4(a)). The externally observed power P_{net} in a frequency-pulling experiment is usually that transmitted past the reflecting discontinuity and is the difference of the power emergent from and incident upon the oscillator (Fig. 7.4(b)).

Analysis of these effects may be carried out with the aid of the circuit shown in Fig. 7.5, in which the load conductance of Fig. 7.1 has been replaced by the transformed output transmission line with characteristic admittance G_L, which is also the load conductance of the free-running oscillator. Standing-wave ratios and the ratio of r.f. power travelling in opposite directions in this transmission line are unaltered by the transformation. Currents i and voltages V in the transmission line are represented in a similar way to eqns (7.1) and (7.2), but losses in the output transmission line are assumed negligible. This is justifiable in the present circumstances because the standing-wave ratios and stored energy are usually small. We have

$$i/G_L = \exp(j\omega t)\{A\exp(-jkx) - D\exp(j\theta)\exp(jkx)\}, \qquad (7.20)$$

$$V = \exp(j\omega t)\{A\exp(-jkx) + D\exp(j\theta)\exp(jkx)\}. \qquad (7.21)$$

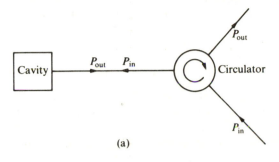

(a)

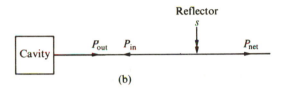

(b)

Fig. 7.4. Schematic diagram of the circuits used for (a) frequency-locking, (b) frequency-pulling

A and D are the real voltage amplitudes of the outgoing and incoming signals, respectively, and θ is the phase difference between them at $x = 0$. If the equivalent circuit of the diode and cavity are placed at $x = 0$, the outgoing power P_{out} and the incoming power P_{in} are

$$P_{out} = \tfrac{1}{2}A^2 G_L, \qquad (7.22)$$

$$P_{in} = \tfrac{1}{2}D^2 G_L, \qquad (7.23)$$

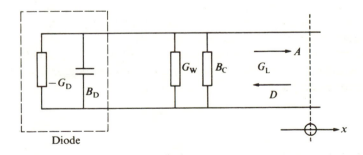

Fig. 7.5. Convenient transmission-line form of the load as used for frequency-locking and frequency-pulling calculations.

83

and

$$P_{net} = P_{out} - P_{in} = \tfrac{1}{2}(A^2 - D^2)G_L .$$ (7.24)

When frequency locking or pulling has been achieved there is only one frequency within the circuit, so eqns (7.20) – (7.23) represent both experimental effects. Writing V_u and i_u for V and i respectively in the absence of a reflecting discontinuity or injected signal, eqns (7.20) and (7.21) become

$$V_u = i_u/G_L = \exp(j\omega t)A_0 \exp(-jkx) .$$ (7.25)

The free-running frequency and r.f. voltage amplitudes are determined by requiring both the total susceptance and conductance to be zero. Frequency-locking and -pulling effects occur when an incoming signal causes the admittance at the input of the transmission line, looking out from the diode and cavity, to differ from G_L. If the admittance perturbation is denoted by $\Delta G_S + j\Delta B_S$ we have

$$\frac{G_L + \Delta G_S + j\Delta B_s}{G_L} = \frac{A^2 - D^2 - 2jAD\sin\theta}{A^2 + D^2 + 2AD\cos\theta} .$$ (7.26)

When $D \ll A$, $A \simeq A_0 \simeq V_0$ and eqn (7.26) may be rewritten to first order as

$$\frac{\Delta G_S + j\Delta B_S}{G_L} = -2\frac{D}{A}\cos\theta - 2j\frac{D}{A}\sin\theta$$

$$\equiv -\frac{2\Delta I}{G_L V_0}\exp(j\theta),$$ (7.27)

where V_0 is the r.f. voltage at the device terminals and ΔI is an equivalent current generator equal to DG_L. It is connected into the equivalent circuit as shown in Fig. 7.6 and sometimes allows convenient simplifications of analysis. ΔG_S and ΔB_S are used in eqns (7.14) and (7.15) or (7.16) and (7.17) to determine the frequency-locking or -pulling characteristics.

7.6.2. Determination of the external Q-factor Q_E

If the frequency locking or pulling is to be used for measurement of Q_E, it is essential that the oscillator conditions are not appreciably altered by the frequency perturbation. An external monitor is needed to identify operating conditions which have the same load conductance G_L and r.f. voltage amplitude as the free-running oscillator. Eqn (7.26) shows that two positions or phases satisfy this criterion and are given by

$$\cos\theta = -D/A$$ (7.28)

and

$$A_0{}^2 = A^2 - D^2 .$$ (7.29)

Eqn (7.29) shows that the frequency shift should be observed at the points where P_{net} is unaltered by the measurement. The susceptance change from free-running conditions using eqn (7.28) in (7.26) is

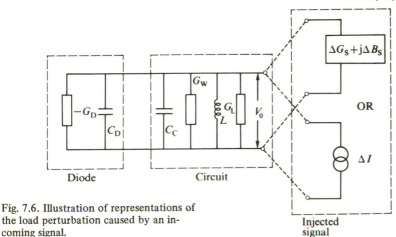

Fig. 7.6. Illustration of representations of the load perturbation caused by an incoming signal.

$$\frac{\Delta B_S}{G_L} = \pm \frac{2}{\{(A/D)^2 - 1\}^{\frac{1}{2}}}.$$ (7.30)

Substitution of ΔB_S in eqn (7.16) with $\Delta G_S = 0$ and using (Montgomery *et al.* 1948) $Q_E = (\omega/2G_L)(\partial B/\partial \omega)$ gives the corresponding change of frequency

$$\Delta \omega_0 = \frac{1}{\{(A/D)^2 - 1\}^{\frac{1}{2}}} \cdot \frac{\omega}{Q_E}.$$ (7.31)

The only approximation involved in this expression is

$$\left(\frac{\partial G}{\partial \omega}\right) \frac{(\partial B/\partial V_0)}{(\partial G/\partial V_0)} \ll \frac{\partial B}{\partial \omega},$$

in the denominator of eqn (7.14). For transferred-electron devices $(\partial B/\partial V_0)/(\partial G/\partial V_0)$ is of order unity (Bestwick *et al.* 1973) and the approximation is usually good to within a few per cent.

In many experimental conditions it will only be possible to identify one frequency at which P_{net} is unaltered, because the other will lie outside the locking or pulling bandwidth. However, the results are usually expressed in terms of the full frequency difference $\Delta \omega_p$ (pulling) or $\Delta \omega_1$ (locking), which equal $2\Delta \omega_0$.

In a locking experiment $(A/D)^2 = P_{out}/P_{in}$, therefore

$$Q_E = \frac{2\omega}{\Delta \omega_1} \cdot \frac{1}{\{(P_{out}/P_{in}) - 1\}^{\frac{1}{2}}} = \frac{2\omega}{\Delta \omega_1}\left(\frac{P_{in}}{P_{out}}\right).$$ (7.32)

The approximation in eqn (7.32) recognizes the usual experimental condition $P_{in} \ll P_{out}$.

In a pulling experiment with voltage–standing-wave ratio s

Circuit properties

$$\frac{A}{D} = \frac{s+1}{s-1} \,,$$

therefore

$$Q_{\mathrm{E}} = \frac{\omega}{\Delta\omega_{\mathrm{p}}} \cdot \frac{(s-1)}{\sqrt{s}} \,. \tag{7.33}$$

This technique of maintaining P_{net} unaltered will suffer sensitivity and accuracy problems when the oscillator is coupled to give near-optimum power output in its free-running state.

7.6.3. Spurious frequency-pulling or -locking interference

If the effect of spurious-load mismatches or interfering signals is to be estimated, the relevant quantity is the complete, locking or pulling, bandwidth. The criterion for determination of the band-edge frequencies is $\partial(\Delta\omega)/\partial\theta = 0$. If the relevant assumption $D \ll A$ is made, we have from eqns (7.16) and (7.17),

$$\left(\frac{\partial B}{\partial\omega}\right)\Delta\omega = \frac{2DG_{\mathrm{L}}}{A}\left\{\sin\theta - \frac{(\partial B/\partial V_0)}{(\partial G/\partial V_0)}\cos\theta\right\}$$

and $\partial(\Delta\omega)/\partial\theta = 0$ when

$$\tan\theta = -\frac{(\partial G/\partial V_0)}{(\partial B/\partial V_0)} \,. \tag{7.34}$$

Using eqn (7.34) and the substitution

$$Q_{\mathrm{E}} = \frac{\omega}{2G_{\mathrm{L}}}\left(\frac{\partial B}{\partial\omega}\right),$$

we have for a locking experiment

$$\Delta\omega_{\mathrm{lf}} = \frac{2\omega}{Q_{\mathrm{E}}}\left(\frac{P_{\mathrm{in}}}{P_{\mathrm{out}}}\right)^{\frac{1}{2}}\left\{1 + \frac{(\partial B/\partial V_0)^2}{(\partial G/\partial V_0)^2}\right\}^{\frac{1}{2}} \tag{7.35}$$

In a pulling experiment

$$\Delta\omega_{\mathrm{pf}} = \frac{2\omega}{Q_{\mathrm{E}}}\frac{(s-1)}{(s+1)}\left\{1 + \frac{(\partial B/\partial V_0)^2}{(\partial G/\partial V_0)^2}\right\}^{\frac{1}{2}} \tag{7.36}$$

The subscript f refers to the full locking or pulling bandwidth. Not only do eqns (7.35) and (7.36) give information on the maximum tolerable mismatches or interfering-signal levels but they also allow estimation of the ratio $(\partial B/\partial V_0)/(\partial G/\partial V_0)$ by comparing the frequency difference to give unaltered P_{net}, with the full locking or pulling bandwidth. An expression often used to calculate Q_{E} or the spurious frequency pulling (Warner and Hobson 1970) is

$$Q_{\mathrm{E}} = \frac{\omega}{2\Delta\omega_{\mathrm{pf}}}\left(s - \frac{1}{s}\right). \tag{7.37}$$

Eqn (7.37) will give the correct order of magnitude but was erroneously derived with the implicit assumption that $(\partial B/\partial V_0) = 0$.

Typical values involved in a frequency-locking experiment would have P_{in} 40 dB below P_{out} at $\omega/2\pi = 10\text{GHz}$ and $Q_E = 100$. If $(\partial B/\partial V_0)/(\partial G/\partial V_0) = \sqrt{3}$, we have

$$\frac{\Delta\omega_{lf}}{2\pi} = 4 \text{ MHz and } \frac{\Delta\omega_l}{2\pi} = 2 \text{ MHz}.$$

If the same oscillator was used in a frequency-pulling experiment with a standing-wave ratio of $1\cdot2$,

$$\frac{\Delta\omega_{pf}}{2\pi} = 36\cdot3 \text{ MHz and } \frac{\Delta\omega_p}{2\pi} = 18\cdot2 \text{ MHz}.$$

7.7. Local modes and defects in frequency tuning

High-stability oscillation has often been achieved with a circuit, illustrated in Fig. 7.7, which stores a large amount of r.f. energy. The diode is mounted centrally on one broad wall of a full-height waveguide and at the end of a post which introduces the d.c. bias and allows r.f. current to couple or flow to the waveguide circuit. In order to store a lot of r.f. energy, the waveguide short-circuit is placed several half wavelengths further from the diode than the minimum necessary to obtain a resonance at the desired frequency.

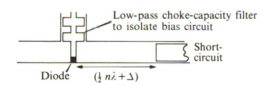

Low-pass choke-capacity filter to isolate bias circuit

Short-circuit

Diode $(\tfrac{1}{2} n\lambda + \Delta)$

Fig. 7.7. A simple multiple-mode full-height waveguide cavity.

Several defects become apparent in the practical operation of this circuit, and they are representative of defects which may occur in any circuit configuration. Oscillation will be possible at frequencies where the circuit resonates with the diode for different values of the integer n at the same setting of the short-circuit, providing that the load conductance reflected to the diode terminals at the appropriate frequency is less than the maximum modulus of the negative conductance of the diode at that frequency. There is also a requirement that $\partial|G_D|/\partial V_0$ is negative for a stable working point. This is normally satisfied if the above conductance criterion is satisfied, owing to the decrease of $|G_D|$ with V_0 at high V_0. The frequency range for appreciable negative conductance is usually an octave or greater, so many frequencies become possible when n is large. When the waveguide short-circuit is moved, all the possible frequencies vary, as do the load

Circuit properties

impedances transformed to the diode terminals. For any one mode (used here and in the literature to describe the relationship between frequency and short-circuit position, with n constant) there are short-circuit positions in which the load conductance transformed to the diode terminals is greater than the maximum modulus of the negative conductance of the diode, so oscillation ceases. Reversal of the short-circuit movement may cause oscillation to start again in that mode, but the oscillation starting position will not coincide with the position where oscillations were quenched. The $|G_D| - V_0$ curve (Fig. 7.8) often has a smaller $|G_D|$ at small V_0 than at the optimum and is one cause of this tuning hysteresis.

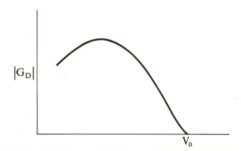

Fig. 7.8. A typical form of the relationship between the r.f. negative conductance and r.f. voltage amplitude.

Frequency-jumping will occur if another mode is not overdamped when the oscillations have been quenched in the first mode. Such effects will complicate the tuning hysteresis, because it will be necessary to reverse the movement of the short-circuit to a point where the loading of the second mode is too great. A typical set of tuning curves for this type of oscillator are shown in Fig. 7.9. If high-stability operation is required at a single frequency, it will be necessary to introduce absorbent material at suitable places within the cavity in order to overload all modes except the desired one. If wide-range mechanical tunability is required, it will be necessary to accept lower stability and work with $n = 1$.

Unfortunately, when $n = 1$ there are still problems which are indicated by the regions in Fig. 7.9 where the frequency is only weakly affected by the position of the waveguide short-circuit. The reason for this behaviour may be explained with the aid of Fig.7.7 and Fig. 7.10. The latter is an axial cross-section of the waveguide mount at right-angles to that in Fig. 7.7. When the short-circuit separation from the diode is comparable with the height of the waveguide an attempt is being made to operate the diode at a frequency where the electrical length of the diode mounting post is comparable with a half wavelength. When viewed along the mounting post the electrical conditions are similar to those of a coaxial mount which has short-circuits at the positions of the top, and bottom, broad-walls of the wave-guide (Taylor, Fray, and Gibbs 1970). Movement of the waveguide short-circuit only alters the impedance of this TEM (*t*ransverse *electro*-

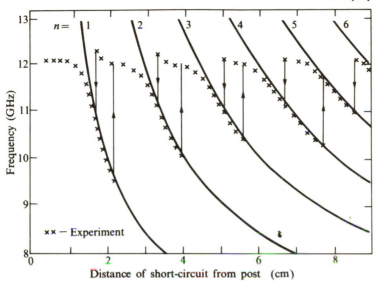

Fig. 7.9. Illustration of multiple-mode operation, frequency-jumping, and hysteresis. Solid lines indicate the cavity length with $\Delta = 0$.

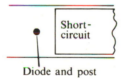

Fig. 7.10. Top view of the circuit in Fig. 7.7.

magnetic) circuit so that it has a very weak effect on the frequency. A similar situation exists when the waveguide short-circuit, in a mode with $n > 1$, reflects an open-circuit to the mounting-post terminals. These 'frequency-saturation' effects may be relieved by reducing the waveguide height or by placing the diode partway up the mounting post (Taylor *et al.* 1970), but the latter technique may carry penalties for the maximum allowable thermal dissipation. Alternatively, it may be better to start afresh with a different type of structure. This has been done by a reversal of the procedure referred to earlier of inserting lossy material into the cavity to damp undesired modes. Instead a cavity was designed (Clemetson, Kenyon, Kurokawa, Owen, and Schlosser 1971) which was overdamped for all modes. This was followed by insertion of a reactive circuit between the diode and lossy material which open-circuited (with a parallel resonant circuit), the latter at one frequency only. A cross-section of this circuit is shown in Fig. 7.11. When the circuits have been 'cleaned', frequency saturation will be set by the parasitic reactances of the diode package (Owens and Cawsey 1970).

Circuit properties

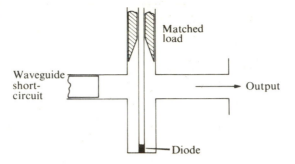

Fig. 7.11. Cavity design by Clemetson, Kenyon, Kurakowa, Owen, and Schlosser (1971).

7.8. Circuit considerations for multi-frequency operation

The previous sections contained implicit assumptions that there was only one steady-state frequency of operation and the harmonic content of the current and voltage waveforms was negligible. This is clearly invalid for high-efficiency oscillators with square-wave or relaxation types of behaviour and is apparently invalid for the device current when operated with a sinusoidal r.f. voltage waveform (see space-charge mode descriptions in Chapter 4). In steady-state operation, the current and voltage waveforms may be Fourier-analysed into their respective harmonics so that the $(n + 1)$-port circuit shown in Fig. 7.12 is a generally valid representation of any free-running oscillator circuit. The boxes marked ω, ..., $n\omega$ are perfect filters which are short-circuits at their contained frequency or open-circuits otherwise. Z_ω .... $Z_{n\omega}$ is the impedance presented to the diode terminals at the subscripted frequency. From Kirchoff's laws we have, in complex terms,

$$i(t) = i_B + \sum_{n=1}^{\infty} i_{n\omega} \exp(jn\omega t) ,$$

$$V(t) = V_B + \sum_{n=1}^{\infty} V_{n\omega} \exp(jn\omega t) .$$

$i_{n\omega}$ and $V_{n\omega}$ are the appropriate complex current and voltage amplitude respectively, related by $Z_{n\omega}$, and i_B and V_B are the corresponding bias quantities. By Fourier analysis of $i(t)$ and $V(t)$ it is possible to equate coefficients at each frequency so that Fig. 7.12 becomes a generalization of the earlier single-frequency equivalent circuits. We cannot now describe the diode by a simple negative conductance and capacity without careful attention to their harmonic frequency-dependencies. When all the n circuit components are capacitive and of low impedance at all harmonics, as is often the case for transit-time oscillators, it becomes possible to represent $Z_{2\omega} ... Z_{n\omega}$ by short-circuits and Z_ω by the reactance considered in previous sections. In practice, the circuit must be designed not to have any inadvertent parallel resonances at frequencies close to the first few harmonics.

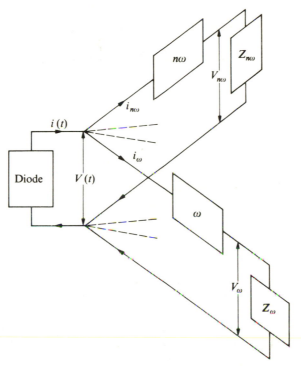

Fig. 7.12. General steady-state equivalent circuit for multiple-harmonic operation.

In such a simplifying case the voltage waveform will be sinusoidal but the current waveform will not, as seen in earlier descriptions of the diode operating modes. However, the relevant admittance of the diode is obtained only from the sinusoidal voltage and the corresponding sinusoidal component of diode current at the fun-

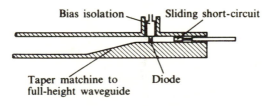

Fig. 7.13. Axial section of a reduced-height waveguide mount.

damental frequency, so that the simple equivalent circuits used in the previous sections are valid. There is no power generation or dissipation at the higher harmonic frequencies of the current owing to the short-circuit impedance presented to the diode by the circuit.

Circuit properties

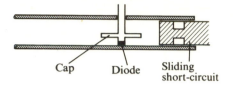

Fig. 7.14. Axial section of a resonant-cap waveguide mount.

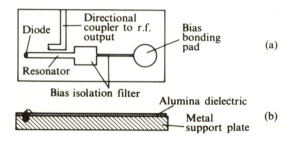

Fig. 7.15. A simple microstrip circuit. The transmission lines are formed between a pattern of top-surface conductors and an 'infinite' ground plane and are separated by alumina dielectric.

7.9. Choice of cavity designs

To conclude the discussion of cavity design a summary of the advantageous and disadvantageous features of several cavity designs mentioned in different parts of the book is given in Table 7.1.

TABLE 7.1

Cavity type	Advantages	Disadvantages
Coaxial (Fig. 8.5)	Medium—low stored energy — easily tunable mechanically and electronically. Simple mechanical construction.	Moderate to poor frequency stability. Tolerable losses in X-band become severe at higher frequencies.
Reduced-height waveguide (fig. 7.13)	Similar to coaxial but construction convenient for matching circuits to amplifiers by diode mount offset from centre. Convenient for mounting tuning varactors in parallel with diode.	Significant internal loss owing to low impedance and consequent high r.f. currents.
Full-height waveguide, diode on central post (Fig. 7.7)	Large stored energy and good frequency stability if frequency-saturation effects are avoided.	Frequency saturation effects and multiple mode of operation. Difficult to tune over wide range.
Full-height waveguide, diode under radial cap on centre post (Fig. 7.14)	Uses controlled frequency saturation, with frequency essentially controlled by cap. Has low internal losses so useful at millimetre-wave frequencies.	Difficult to tune and match to optimum load.
Kurokawa circuit (Fig. 7.11)	Reduction of number of modes compared with full-height waveguide and control of spurious out-of-band modes.	Mechanically inflexible and complicated. Difficult to tune.
Lumped circuit (Fig. 8.6)	Very low stored energy — useful for extreme tuning ranges and relaxation oscillation.	Large circuit losses.
Microstrip (Fig. 7.15)	Amenable to large scale production of integrated microwave systems.	Large circuit losses. Difficult to tune.
Conical and Biconical circuits (Fig. 4.10)	Similar to coaxial but smaller mounting parasitic reactances. Useful for low-loss 'lumped' circuits in relaxation modes and for device characterization.	Mechanically cumbersome and difficult to tune.

8.1. Frequency modulation

Frequency modulation may be achieved by control of any susceptive elements in the simple equivalent circuit of a transferred-electron oscillator and its controlling circuit, which are illustrated in Fig. 8.1. Bias-circuit modulation is achieved through control of the device susceptance B_D. Magnetic tuning is obtained by

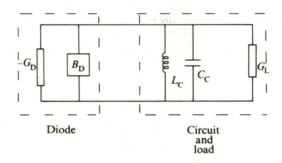

Diode

Circuit
and
load

Fig. 8.1. Equivalent circuit of device and cavity.

strongly coupling the magnetic field of L_C into the ferromagnetic spin resonance of yttrium iron garnet or a similar material. Varactor tuning utilizes control of C_C. In each case we may expect that the greatest frequency variation will be obtained when the appropriate circuit element has a large or dominant storage of microwave energy.

8.2. Varactor tuning

The circuit capacity C_C in Fig. 8.1 may be realized with a variable capacitance (varactor) diode to give frequency control. The oscillator diode, varactor, and circuits, whose simple equivalent circuits are illustrated in Fig. 8.2, may be combined in the series or parallel forms of Fig. 8.3. In practice, each component has a Q-factor much greater than unity, allowing the approximate equivalences which have been used in Fig. 8.2. The loaded Q-factor of the entire circuit Q_L, the oscillator-diode Qfactor Q_G, and the varactor Q-factor Q_v are defined in

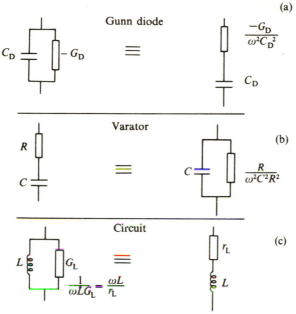

Fig. 8.2. Circuit equivalences for varactor tuning:

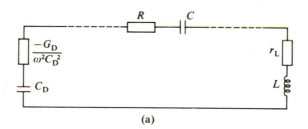

(a)

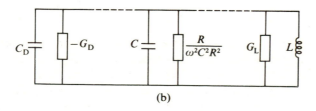

(b)

Fig. 8.3. (a) series and (b) parallel mounting of a varactor.

95

Desirable modulation

convenient forms in the following equations for each circuit configuration:

Parallel	Series

$$Q_L = \frac{1}{\omega_0 L G_D} \qquad\qquad\qquad Q_L = \frac{\omega_0 L}{(r_L + R)}$$

$$Q_G = \frac{\omega_0 G_D}{G_D} \qquad\qquad\qquad Q_G = \frac{\omega_0 C_D}{G_D}$$

$$Q_V = \frac{1}{\omega_0 R C_0} \qquad\qquad\qquad Q_v = \frac{1}{\omega_0 R C_0}$$

$$\omega^2 = \frac{1}{L(C + C_D)} \qquad\qquad\qquad \omega^2 = \frac{(C + C_D)}{L C C_D}$$

$$\frac{P_v}{P_{tot}} = \frac{\omega^2 C^2 R}{G_D} \qquad\qquad\qquad \frac{P_v}{P_{tot}} = \frac{\omega^2 C_D{}^2 R}{G_D} = \frac{R}{(r_L + R)}.$$

ω_0 is the centre frequency at which $C = C_0$ and r_L is equal to $\omega_0{}^2 L^2 G_L$. P_v is the r.f. power dissipated in the spurious series resistance of the varactor, and P_{tot} is the total r.f. power generated by the oscillator diode. If a tuning figure of merit M is defined as $((\Delta\omega/\omega_0)/(\Delta C/C_0))$ analysis of each circuit configuration shows

Parallel	Series

$$M_p = \frac{-C_{op}}{2(C_{op} + C_D)} \qquad\qquad\qquad M_s = \frac{-C_D}{2(C_{os} + C_D)}$$

$$\frac{P_v}{P_{tot}} = \frac{Q_G}{Q_v} \cdot \frac{C_{op}}{C_D} \qquad\qquad\qquad \frac{P_v}{P_{tot}} = \frac{Q_G}{Q_v} \cdot \frac{C_D}{C_{os}}$$

$$Q_L = Q_G \left(\frac{C_{op} + C_D}{C_D}\right) \qquad\qquad\qquad Q_L = Q_G \left(\frac{C_{os} + C_D}{C_{os}}\right)$$

$$\Delta\omega = -\frac{\Delta C}{C_{op}} \cdot f_{cv} \cdot \frac{1}{2Q_L} \cdot \frac{P_v}{P_{tot}} \qquad\qquad \Delta\omega = \frac{-\Delta C}{C_{os}} \cdot f_{cv} \cdot \frac{1}{2Q_L} \cdot \frac{P_v}{P_{tot}},$$

where $f_{cv} = 1/RC_0$ is the cut-off frequency of the varactor, C_{op} is the capacity of the varactor chosen for parallel circuit control, and C_{os} is that for the series circuit at the mean varactor bias voltage. From the above identities it can be seen that the two configurations have identical behaviour if $C_{op}/C_D = C_D/C_{os}$. The only difference between the two configurations is the parallel-circuit requirement for a capacity which is appreciably greater than the oscillator-diode capacity, and vice versa for the series circuit. When $C_{op} \gg C_D$ and $C_{os} \ll C_D$, M_s and M_p approach the limiting value of 0·5 which is appropriate to a simple LC circuit in which one element is controllable. There is a limit to the ratios C_{op}/C_D and

C_D/C_{0s} owing to loss of r.f. power in the spurious varactor resistance. As the varactor reactance is made dominant so its resistance also becomes dominant. It is necessary to use varactors with $Q_v \gg Q_G$ in order to avoid appreciable power loss without excessive loss of tuning range $\Delta\omega_T$. A good varactor can be seen to be one with optimum $(\Delta C/C)_{max} \cdot f_{cv}$ (Cawsey 1970). The other components in the expression for $\Delta\omega$ are circuit constraints outside the varactor. The minimum allowable Q_L is dictated by noise and stability requirements. A varactor with a cut-off frequency of 100 GHz and $\Delta C/C_0 = 1$ mounted in a circuit with $Q_L = 50$ in which half the r.f. power is lost in the varactor would have $\Delta\omega_T/2\pi = 0\cdot5$ $2\pi/$GHz. Great care is needed in circuit design to reduce Q_L below 50 and to achieve $\Delta C/C_0 = 1$ owing to stray mounting capacity around the varactor. A large-signal restriction on $\Delta C/C_0$ arises from the desirability of avoiding excessive varactor losses when the r.f. power, coupled into the varactor, instantaneously swings it into forward conduction or reverse avalanche breakdown. Varactors are now available with cut-off frequencies ~ 1000 GHz (with the appropriate price disadvantage), and tuning ranges between 1 GHz and 4 GHz have been achieved (M.E.S.L. 1972; Downing and Myers 1971; Corbey and Gough 1973). Practical circuits do not usually resemble the outline circuits of Fig. 8.3. In general, the varactor will be reactively coupled to the circuit by an impedance transformation such as illustrated in Fig. 8.4. When the ultimate tuning ranges are not required a convenient circuit-mounting is illustrated in Fig. 8.5, where the transformer is the loop coupling of the varactor into the coaxial cavity.

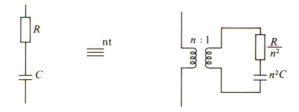

Fig. 8.4. Transformed coupling of varactor equivalent circuit.

One of the attractive features of varactor tuning is its fast response. Tuning rates up to 1 GHz μs^{-1} have been achieved. A disadvantage is the non-linear relationship between the capacity and the bias voltage. Linearizing circuits are required for some applications with cost and reliability disadvantages and the further possibility that they may restrict the tuning rate. Care has to be taken with heat-sinking of the varactor diode to avoid slow 'settling' times after a step-function change of bias voltage. The change of r.f. power dissipation may cause a significant thermal variation of the varactor capacity.

Desirable modulation

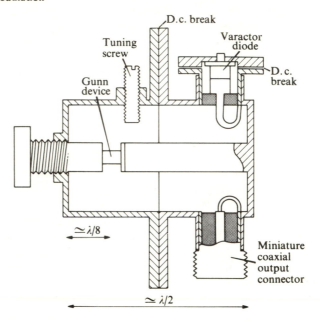

Fig. 8.5. A coaxial X-band Gunn oscillator cavity (Wilson 1969).

8.3. YIG tuning

Inductive-circuit tuning of c.w. oscillators can be created by magnetically coupling the r.f. current of an oscillator diode to the electron spins in a sphere of *y*ttrium *i*ron *g*arnet (YIG) so that the frequency is locked to the ferromagnetic resonance. A typical circuit illustrating the coupling between the diode, its YIG 'cavity', and the external circuit is shown in Fig. 8.6. Linear frequency control (Fig. 8.7) is obtained by variation of an external magnetic field $H(\text{A m}^{-1})$, according to the relationship

$$f = 3{\cdot}5 \times 10^4 \times H \text{ Hz.}$$

Linearity (ratio of frequency error to centre frequency) can be better than 10^{-3} (L.E.P. 1972), but both linearity and hysteresis are primarily functions of the magnetic driving components which usually restrict the sweep rate to 1 MHz μs^{-1}. If the YIG sphere is orientated with a suitable (not high-symmetry) crystallographic direction parallel to the magnetic field, it's resonant frequency is independent of temperature (Easson 1971). The unloaded Q-factor of the resonance is $\sim 10\,000$, but tight coupling to the Gunn diode will usually result in a loaded Q-factor of ~ 100. Temperature stability is typically 1 MHz K^{-1}, and the oscillator linewidth is a few kilohertz (Easson 1971; L.E.P. 1972). This performance is satisfactory for many applications, such as $C-$, $X-$, and $J-$ band swept-frequency oscillators.

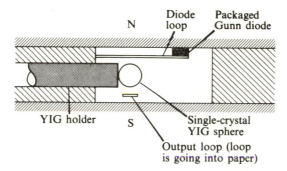

Fig. 8.6. A Gunn oscillator tuned with a YIG sphere (Omori 1969).

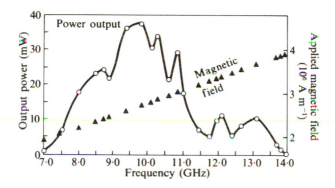

Fig. 8.7. Tuning response of a YIG-tuned oscillator (Omori 1969).

The solid-state reliability ($> 30\,000$ hours life (Coultas 1972)) is considerably better than that of the backward-wave oscillators they have replaced (life $\sim$ 1000 hours). The maximum output power appears to be limited somewhere in the range 10 mW to 100 mW by saturation effects in the YIG.

8.4. Bias-circuit frequency modulation

Modulation by bias-voltage variation is usually only used for small and empirical frequency adjustments (< 1 percent), but knowledge of it is necessary for power-supply design to achieve any system-stability specification. The modulation co-efficient m_f (incremental ratio of frequency and bias voltage) is dependent on the modulation frequency f_m, on the diode and its operating conditions. At modulation frequencies below 1 MHz, c.w. oscillators at X-band or above have a dependence of m_f on f_m which is partly caused by thermal relaxation processes. In essence, the instantaneous frequency is related to the instantaneous device temperature through the frequency–temperature relationship of the diode

Desirable modulation

(De Sa and Hobson 1971). The effect is greatest in diodes showing both the greatest contact resistance and frequency variation with ambient temperature. The frequency and phase of the modulation are predicted well by the thermal relaxation discussed in § 6.3.2. Fig. 8.8 illustrates the form of the variation of m_f with f_m for a diode with poor frequency stability against changes of ambient temperature. As f_m tends to zero the instantaneous frequency decreases with an increase of bias voltage for this type of diode. $|m_f|$ is usually less than 50 MHz V^{-1} for an X-band c.w. oscillator with $Q_L = 100$ when coupled for optimum power output.

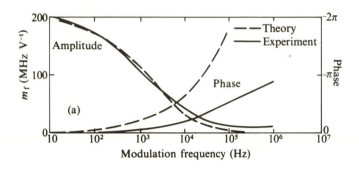

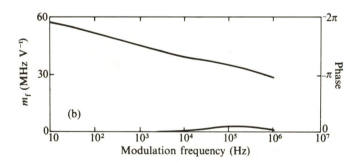

Fig. 8.8. (a) Variation of the magnitude and phase of the frequency deviation with modulating frequency for bias-voltage modulation. (b) A typical frequency-deviation characteristic for a diode that exhibits weaker modulation properties. The phase variation is correspondingly reduced (De Sa and Hobson 1971).

In the absence of thermal modulation effects, bias-circuit modulation arises from the 'dynamic susceptance' of the oscillator and is independent of f_m. The effect may be understood by reference to Fig. 8.9 for the delay mode but will occur in any device with a hysteresis in its resistive non-linearity. The current waveform may be Fourier-analysed into a component i_a in antiphase with the r.f. voltage and one i_q in quadrature (capacitive). Over the small frequency ranges

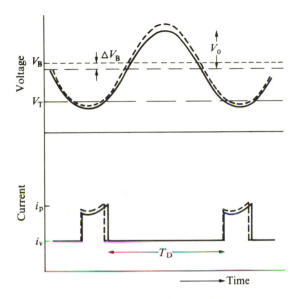

Fig. 8.9. Illustration of the change of r.f. voltage and current with bias voltage.

under consideration ($\sim$ 1 percent), the load conductance G_L reflected to the active diode terminals from the external circuit will remain essentially constant, i.e.

$$G_L = |i_a/V_0|. \tag{8.1}$$

The quadrature current causes a susceptance B_q where

$$B_q = i_q/V_0. \tag{8.2}$$

An alteration of bias voltage will change the domain transit-time T_D and the magnitudes of V_0 and i_a, but the condition in eqn (8.1) must remain satisfied. Both i_q and V_0 change so that the susceptance will also change and cause a frequency alteration in order to maintain a conjugate match between diode and circuit. There will also be a temperature effect in this process owing to temperature-dependence of both the velocity–field characteristic and T_D. This component will show the characteristic thermal relationship between f_m and m_f but is much less pronounced in devices which have good frequency-stability against ambient temperature changes. Their instantaneous frequency usually increases with bias voltage, and m_f is often less than 50 MHz V^{-1} for X-band c.w. oscillators with the above cavity conditions. The variation of m_f with temperature at quasi-d.c. modulation frequency is illustrated in Fig. 8.10 when both the thermal-relaxation and dynamic-susceptance processes have approximately equal magnitudes. Only the former is strongly temperature-dependent. Thermal relaxation dominates at low temperature (where $\mathrm{d}f/\mathrm{d}T$ is large) with a negative m_f, but there is a positive m_f at the higher temperatures.

Desirable modulation

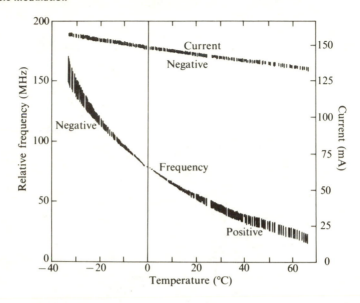

Fig. 8.10. Oscillator frequency and current as a function of temperature. The width of each trace is the peak-to-peak modulation of the relevant quantity caused by the bias-voltage modulation. The positive (in-phase) and negative (antiphase) captions refer to the phase of the modulation relative to that of the modulating voltage (De Sa and Hobson 1971).

In both of the above processes there has been an implicit assumption that the r.f. currents and voltages are in quasi-static equilibrium with the modulation voltage. This assumption is not true for $f_m \gtrsim 10$ MHz and a new dominant modulation process occurs. The modulation process is a result of the particular equivalent circuit of a transferred-electron device as illustrated in Fig. 8.11. For an X-band (or higher-frequency) c.w. oscillator the capacitive susceptance is much greater than the magnitude of the negative conductance, so that the relationship of cavity voltage and current at the diode is dominated by the capacitor.

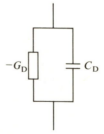

Fig. 8.11. Simplest equivalent circuit of a transferred-electron device.

Application of a step-change of bias voltage causes an unbalance between the diode negative conductance and the load conductance reflected to the diode terminals, so that growth or decay of the r.f. voltage amplitude V_0 occurs at a rate limited by the stored r.f. energy in the oscillator cavity. The extra power generation or dissipation occurs in the device conductance, and so it necessarily occurs in quadrature with the capacitive current and the cavity controlling current. This quadrature power injection causes a phase modulation in sympathy with the amplitude modulation implied by the changing r.f. voltage amplitude. The modulation lasts for the duration of the r.f. voltage readjustment. Transformation into the frequency domain yields (Hobson, Kocabiyikoglu, and Martin 1970)

$$-\frac{1}{5}\pi\Delta\theta \approx \frac{\Delta V_0}{V_0} = \frac{\Delta V_{B0}}{V_0} \cdot \frac{(\partial G_D/\partial V_B)}{(\partial G_D/\partial V_0)} \cdot \exp(j\omega_m t)$$

when

$$\omega_m < \frac{V_0}{4\alpha}\left(\frac{\partial G_D}{\partial V_0}\right),$$

for a bias-voltage modulation of

$$\Delta V_B = \Delta V_{B0} \exp(j\omega_m t).$$

$\Delta\phi$ is the phase modulation, ΔV_0 is the amplitude modulation, G_D is the device conductance-and α is a constant related to the r.f. energy stored in the cavity.

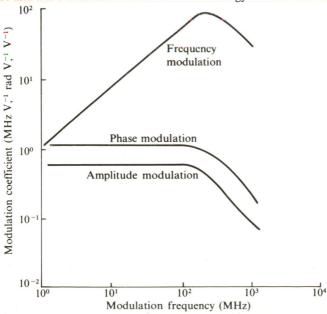

Fig. 8.12. High-speed modulation coefficients of a Gunn oscillator (Hobson, Kocabiyikoglu, and Martin 1970).

Desirable modulation

Fig. 8.12 shows a typical relationship between the angle or amplitude modulation and the modulation frequency. The angle modulation is expressed as a phase deviation $\Delta\phi$ or as an equivalent frequency deviation $\Delta\phi f_m$. The upper limit to the modulation-frequency response occurs when the r.f. voltage amplitude does not have time to show an appreciable change in one modulation cycle.

8.5. Bias-circuit amplitude modulation

Under c.w. conditions increasing the bias voltage causes the r.f. power to increase from zero, pass through a maximum, and fall again. Under pulse-bias conditions with no device heating, the r.f. power will often increase monotonically with increasing bias voltage until avalanche breakdown occurs. These effects are illustrated in Fig. 8.13. If linear amplitude modulation is attempted by bias modulation it is likely to be disappointing because the curves in Fig. 8.13 imply a voltage-dependence of the magnitude of the modulation coefficient and of it's sign. There will be a modulation frequency-dependence which will follow the thermal response described earlier for frequency modulation. An improvement in fidelity may be achieved with envelope feedback in which a portion of the modulation signal is demodulated and used in a negative-feedback loop to correct modulation non-linearity (Warner, private communication; Bulman *et al.* 1972).

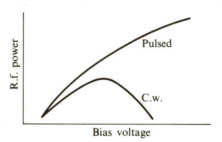

Fig. 8.13. Comparison of the typical variations with bias voltage of the r.f. output power of pulsed and c.w. cavity-controlled transit-time oscillators.

Pulse-width or pulse-code amplitude modulation, where the r.f. power is switched on and off, are more suitable for bias-circuit modulation of transferred-electron oscillators. They are eminently suitable for the generation of narrow pulses for high-resolution radars or high bit-rate modulation, because the rise time can easily be made ~ 1 ns. Assuming that the diode is mounted in a circuit with loaded Q-factor of approximately 50, and the negative conductance under small-signal conditions at the start of oscillation is larger than, but comparable with, the steady-state negative conductance, we see that the rise time of an X-band oscillator ($\sim Q/2\pi$ cycles) is approximately 8 r.f. cycles or 0.8 ns. A fuller description of the starting process and effects which can slow it down are given elsewhere (Kocabiyikoglu and Hobson 1973). A typical start of oscillation is illustrated in Fig. 8.14.

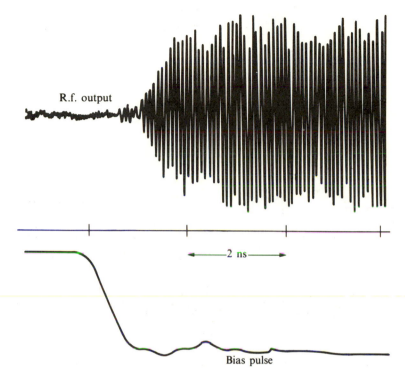

Fig. 8.14. Illustration of the rapid start of oscillation of an X-band Gunn oscillator in a cavity with a loaded $Q \simeq 50$, when coupled for optimum steady-state power output.

8.6. PIN-diode modulation

It is possible to avoid mixing modulation problems with oscillator-stability problems by externally modulating the r.f. output of a c.w. unmodulated oscillator. Amplitude modulation, either linear or switched, can be achieved with a PIN diode used as a voltage variable attenuator in a transmission circuit. Digital-phase modulation can be realized by the circuit illustrated in Fig. 8.15. The PIN diode is placed a quarter of a wavelength in front of a short-circuit and switched between an open-circuit and short-circuit condition. There is a signal-path difference of half a wavelength between reflection from the PIN diode and reflection from the short-circuit, so digital-phase inversion can be achieved. The circulator is necessary to convert the reflection circuit to a transmission modulator.

105

Desirable modulation

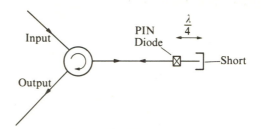

Fig. 8.15. A PIN diode digital-phase modulation circuit.

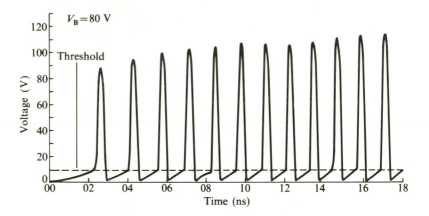

Fig. 8.16. The starting waveform of an LSA-relaxation oscillator (Jeppesen and Jeppsson 1971).

8.7. Bias modulation of relaxation oscillators

The nature of LSA-relaxation oscillators (§ 4.10) causes them to start oscillation within approximately 1 period of exceeding threshold (Fig. 8.16), and they are suitable for high-resolution radar applications. Once operating, they have a characteristically large increase of frequency with bias voltage (Fig. 8.17). That section of their voltage waveform which is rising towards threshold becomes shorter as the bias voltage is increased, because the bias voltage is the asymptote of the exponentially varying voltage. If we make an oversimplified assumption that this section of the waveform starts at half the threshold voltage $\frac{1}{2}V_T$, and the frequency is equal to the inverse of the sum of the time spent below threshold and a constant time $1/f_{as}$ above threshold, the frequency f will be related to the bias voltage V_B by

$$f\tau = 1 \left/ \ln\left\{\frac{V_B - \frac{1}{2}V_T}{V_B - V_T}\right\} + \frac{1}{f_{as}\tau} \right. .$$

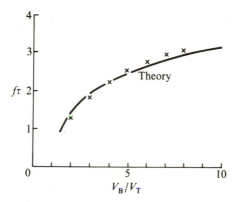

Fig. 8.17. Bias-voltage tuning of an LSA-relaxation oscillator (Jeppesen and Jeppsson 1971).

τ is the low-field relaxation time of the simple series LR circuit used earlier in the description of the relaxation oscillator, and f_{as} is the asymptotic frequency at infinite bias voltage. This expression is compared with a measured variation in Fig. 8.17 with $f_{as}\tau = 3\cdot5$.

9 | Undesirable modulation

9.1. Oscillator noise

For most applications amplitude modulation (a.m.) and frequency modulation (f.m.) noise are not excessively troublesome in transferred-electron oscillators, and so these forms of noise are imperfectly understood. It is known that low-frequency current noise in the bias circuit is partially correlated with f.m. noise (Faulkner and Meade 1968) and there is a partial correlation of the f.m. noise with a.m. noise (Hashiguchi and Okoshi 1971). Both types of oscillator noise (Fig. 9.1) have a characteristic decrease of noise power (or frequency deviation) with increasing modulation frequency, which suggests that their origin is in 'flicker' or generation–recombination processes (De Caqueray, Blasquez, and Graffeuil 1972). Further discussion of the physical background and explanation of the units of f.m. noise measurement are given by Bulman *et al.* (1972).

The magnitude of the noise fluctuations may be minimized by mounting the diode in a cavity with a large r.f. energy storage (see § 7.4). The performance illustrated in Fig. 9.1 was obtained with such a cavity. The decrease of noise power with increasing modulation frequency is of particular value for local oscillators in superheterodyne receivers in which the only troublesome noise is that occurring at a high modulation frequency equal to the intermediate frequency of the receiver.

9.2. Frequency variation with ambient temperature

The variation of centre frequency with the ambient temperature (the $f-T$ relationship) is a serious stability problem for communications systems, both for channel allocation and correct frequency-tracking of remote transmitters and receivers. Stability may be provided with a secondary, high-Q cavity with a large r.f. energy storage (Ito, Komizo, and Sasagawa 1970; Wilson, Tebby, and Langdon 1971), but considerable improvement may often be obtained with good device-technology. These two approaches are typical of any stability problem. On the one hand the diode generates phase or frequency fluctuations, while the circuit provides a smoothing 'flywheel'. Attention to either detail can improve stability (see § 7.4).

Two types of $f-T$ behaviour are commonly observed in c.w. oscillators. Devices containing a metal cathode contact often exhibit a negative df/dT whose

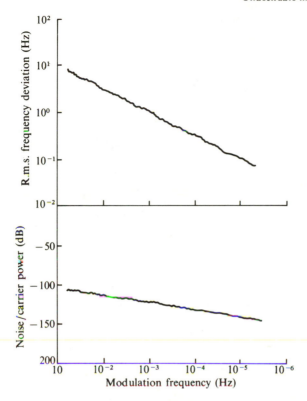

Fig. 9.1. Typical frequency-modulation and amplitude-modulation noise spectra for a c.w. Gunn oscillator mounted in a cavity with large stored energy ($Q_L \simeq 1000$ at full power output). The noise is the sum of that in the two sidebands symmetrically displaced about the centre frequency and is normalized to a bandwidth of 1 Hz.

magnitude decreases with increasing temperature and bias voltage (Bird, Bolton, Edridge, De Sa, and Hobson 1971; Hobson 1972a). The df/dT (Fig. 9.2) is typically between 1 MHz and 10 MHz K^{-1} at room temperature for an X-band oscillator with a loaded Q-factor of 100 at full power output. Communications systems usually require an over-all stability of 1 MHz in X-band over all possible conditions. The large df/DT appears to be caused by a temperature-dependent contact resistance (see § 6.2) and its parallel capacity. They cause a frequency-dependent transformation of the circuit susceptance to the ends of the active region. In Fig. 9.3 the contact damage region is represented by R_C in parallel with C_C. They are a series perturbation of the active device region which is represented by $-G_D$ and C_D. The device must be a conjugate match to the circuit, and the temperature-dependence enters through the temperature-dependence of R_C. Expirical evidence of the temperature-dependence of R_C (Bolton and Jones 1969) may be expressed in the form

109

Undesirable modulation

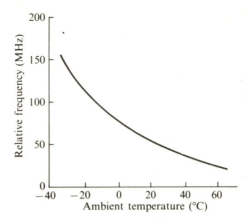

Fig. 9.2. A typical variation of centre frequency with ambient temperature for a c.w. Gunn oscillator with a poor-quality metal alloy contact at its cathode. The circuit had $Q_L \simeq 50$ at full power output.

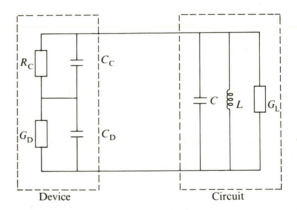

Fig. 9.3. Inclusion of the contact equivalent circuit in that of the device and circuit.

$$R_C \simeq \exp\left(T_C/T\right).$$

T_C is a constant or activation temperature (presumably for electrons from trapping centres or a similar process) which usually lies between 400 K and 1200 K. A simple qualitative description of the $f-T$ variation follows by neglecting G_D, which carries appreciably smaller current than C_D for X-band operation. At low temperatures R_C is essentially open-circuit, so the series combination of C_C and C_D (i.e. a smaller capacity than C_D) must resonate with the circuit. At high temperatures R_C becomes small, so C_D alone must resonate with the circuit. Accordingly, the frequency falls with the higher effective device capacity at the higher temperature. A more detailed account (Hobson 1972a) shows a satisfactory

110

explanation of the $f-T$ behaviour for this type of device.

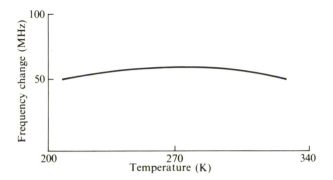

Fig. 9.4. High-stability behaviour against ambient-temperature changes for a device with a good-quality n⁺ cathode contact. Cavity conditions as Fig. 9.2.

Good device-technology should remove the above mechanism of frequency variation, and devices with epitaxial n⁺ contact layers often show a considerably better frequency stability (Bird *et al.* 1971). The $f-T$ curve (Fig. 9.4) is essentially an 'upturned bowl'. An over-all frequency variation less than 10 MHz over the range 200 K to 340 K may be obtained under the same circuit conditions as described for metal-contact devices. Up to 3 dB output-power decoupling may be required. The operating conditions are restricted, otherwise a nearly-linear, negative df/dT occurs with less stability. The operating conditions appear to be the same as required for near-optimum power adjustment of the delayed-domain mode under the incipient space-charge conditions described in § 4.12 (De Sa and Hobson 1973–4). The frequency variation is caused by the detailed temperature-dependence of the dynamic susceptance (described in § 8.4 on modulation) which remains after removal of gross defects like contact resistance. The dynamic susceptance varies owing to its dependence on the temperature-dependent space-charge transit-time and shape of the $v-E$ characteristic.

It must be emphasized that the division between the two device types is blurred. Poor stability is obtainable from badly prepared n⁺ devices or good stability may be found in metal-contact devices when, presumably, they are well prepared.

The above discussion of $f-T$ relationships has been concerned with c.w. devices. High-power LSA-relaxation oscillators also have an $f-T$ problem. The inverse temperature-dependence of their low-field resistance alters the time constant of the exponential rise of voltage below threshold (§ 8.7). An ambient-temperature sensor may be used to control the bias voltage in order to compensate this effect through the strong bias-voltage dependence of the oscillation frequency (Wasse *et al.* 1972).

Undesirable modulation
9.3. Starting-delay time of pulsed oscillators

It is convenient to follow a discussion of the effect of contact resistance on $f-T$ relationships by an outline of excessive starting delays in pulsed X-band transit-time oscillators. Very poor devices may exhibit a delay of milliseconds or longer before device-heating sufficiently reduces the contact resistance to allow a net negative conductance in the circuit. Similar effects may cause oscillations to cease or be unreliable at low temperature. When the contact resistance is not large enough to damp the circuit there is a delay, reckoned in nanoseconds, in circuits designed to allow rapid start of oscillation. Such circuits must store little r.f. energy. An excessive delay is caused by a reduction of the signal growth rate caused by the spurious and dissipative contact resistance (Kocabiyikoglu and Hobson 1973). The starting time correlates well with the apparent contact resistance derived from excess threshold voltage measurements (Edwards, Kellett, and Myers 1972), and it's temperature-dependence is simply related to the $f-T$ relationship of c.w. oscillators (De Sa, Kocabiyikoglu, and Hobson 1971).

QUOT HOMINES,TOT SENTENTIAE

10.1. Indium phosphide

The general principles of transferred-electron devices have been outlined in the previous chapters by reference to the properties of GaAs. Even though Gunn reported the same type of oscillatory behaviour in InP devices when he announced his effect, little work was carried out on this material owing to the relatively poor state of it's device-technology. The prediction that InP should have somewhat different properties to GaAs because it had two sets of satellite valleys (Hilsum and Rees 1970) stimulated further interest. Infrequent or weak scattering was supposed to occur between the lower ($\langle 111 \rangle$) satellite valleys and the central valley, whereas frequent or strong scattering was supposed to occur both between the upper ($\langle 100 \rangle$) satellite valleys and the central valley and between the two sets of satellite valleys. These effects were expected to slow the cyclic intervalley transfer of electrons which is required in any device operation. A high effective diffusion coefficient would result, as described in § 2.10, and it was thought, for example, that LSA operation may be more easily achieved by the reduction of space-charge growth rate. It seems reasonable to expect a reduction in the upper frequency-limit of operation as a penalty if this diffusion mechanism is strong, because a negative-differential conductance requires electron transfer. However, the upper frequency-limit of InP devices does not appear to be smaller than that of GaAs devices (Colliver and Prew 1972).

Reliable measurements are hampered by the material technology of InP, so there is considerable debate about the quantitative form of the $v-E$ character-istic and it's relationship to two- or three-level operation (Hilsum and Rees 1972; Boers 1973). The threshold field appears to be somewhere in the range $0 \cdot 6$ kV mm^{-1} to $1 \cdot 2$ kV mm^{-1} with $1 \cdot 0$ kV mm^{-1} as the most probable value. Measure-ments and predictions of the peak—valley ratio of the $v-E$ characteristic lie between $2 \cdot 6$ and 4, with $3 \cdot 2$ as the most probable value. If the promise of easy space-charge control in LSA devices is not fulfilled the higher peak—valley ratio of InP may still be attractive for increasing the efficiency of LSA-relaxation oscillators. One high-efficiency result has been reported (Wood, Tree, and Paxman 1972), with many characteristics of relaxation oscillators.

Another mode of oscillation giving high efficiencies (up to 20 per cent appears to be related to non-ohmic cathode contacts, but such devices are difficult to manufacture reproducibly. Yu, Tantraporn, and Young (1971) have predicted that the appropriate non-ohmic cathode boundary conditions may allow oscillation with greater efficiency than is the case with ohmic boundaries. The practical difficulty of making ohmic contacts to InP may accidentally provide the required boundary conditions for high efficiency in some cases.

The other area of promise with InP lies in the good noise performance of supercritical amplifiers, as described in § 5.4.

10.2. Non-linear mixing

Gunn oscillators are often used as local oscillators for mixers in microwave receivers. Owing to the non-linearities inherent in the oscillating device it may also take the part of the mixer diode and be used as a self-oscillating mixer (Albrecht and Bechteler, 1970; Lazarus, Bullimore, and Novak 1971). Furthermore, gain may be obtained at the intermediate frequency owing to the END conductance described in § 5.6, so that some part of the intermediate-frequency amplification may also be carried out in the Gunn oscillator. Unfortunately, the noise figure is usually $\sim$ 20 dB, so this compact device does not find application in high-quality receivers, but it may be of value when cheapness is of more value than sensitivity.

Another demodulation process occurs in Gunn oscillators which are frequency locked to an incoming signal (Bestwick *et al.* 1973). If the incoming signal is frequency modulated there is a sympathetic variation of bias current when the Gunn oscillator is run under constant voltage bias. This mechanism may be used to directly demodulate frequency-modulated microwave signals without the intervention of a conventional discriminator. The demodulated signal is substantially independent of the incoming r.f. power, providing that a frequency lock is maintained, so there is an inherent automatic gain-control mechanism. The device acts as a combined limiter, discriminator, and detector by comparison with a conventional f.m. demodulator. Explanation of the effect may be made with a first-order non-linear theory following the same lines as that in § § 7.4 and 7.6. The demodulation process is sensitive in the sense that demodulated signal levels from a 1 MHz frequency deviation usually lie between 1 mV and 10 mV and can be seen on a typical laboratory oscilloscope.

10.3. Logic and functional devices

Reference to Fig. 2.7 shows that a Gunn device biased just below the threshold voltage may carry one of two possible currents, depending on the presence or absence of a domain. This property offers possibilities for high-speed binary logic. Control of domain triggering in uniform or non-uniform devices with two or three contacts allows the realization of many types of logic device (Hartnagel and Izadpanah 1968; Bulman *et al.* 1972). One of the principle difficulties in implementing systems-use of these devices lies in the considerable power dissipation in each unit for both binary states. A minimum power dissipation is implied by the $N_D l$ and $N_D d$ limitations for instability and mature domain formation.

When a stable domain is present and propagating through a device it requires a nearly-constant electric field outside the domain irrespective of the device uniformity, providing that variations occur over a scale larger than required for domain readjustment. Consequently, the domain maintains a nearly-constant electron-drift velocity just outside the domain but the total device current will respond in sympathy to changes in lateral dimensions, doping, or mobility at the position of the domain. Shoji (1967) has used these effects to realize a variety of high-speed waveform generators some of which may be controlled from cycle to cycle by external switching (Bulman *et al.* 1972).

10.4. Microwave systems

The lost cost, light weight, and convenient power supply requirements of transferred-electron devices are allowing the implementation of various microwave systems which were previously impractical outside a military enviroment (Bulman *et al.* 1972). Low-power sources (~ 10 mW) at frequencies around 10 GHz have found the greatest application. They have been used predominantly as fixed-frequency oscillators in Doppler radar systems for intruder detection (Giles and Saw 1972). Complete combinations of the device and cavity are presently marketed by several manufacturers. Klystrons have been replaced as local oscillators in many communications and radar systems by devices similar to the above ones, except for the addition of some mechanical and electronic tuning. Particular advantage has been found in light-weight portable systems (Bulman *et al.* 1972). Other applications are found in laboratory signal sources and in microwave instrumentation for process control (Coultas 1972). Several useful operating details of these signal sources have been described by Wilson (1969).

Higher-power c.w. oscillators do not attract much attention until they exceed the 1 W barrier. Such devices are of interest in communications systems, but this power is difficult to achieve with a single device. Instead, the power output of several diodes must be combined, and an oscillator containing four devices is commercially available. The combination of many devices in a single-frequency oscillator presents many problems, owing to the several modes of interaction between each device.

Short-range, high-resolution radars have been built by exploiting the very rapid start of oscillation described in § 8.5. One example of such developments is seen in a docking aid for large oil-tankers. A continuous indication is given of position and of velocities which are difficult to see with the naked eye.

10.5. Conclusion

At the time of writing, work is proceeding on the exploitation of Gunn devices in many enviroments which previously could not afford the luxury of a micro-

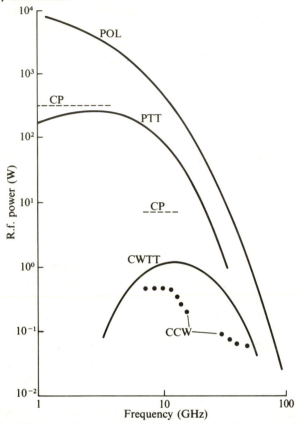

Fig. 10.1. State-of-the-art performance in r.f. power generation by transferred-electron devices (1972–73). Best results: POL – pulsed overlength devices; PTT – pulsed transit-time devices; CWTT – c.w. transit-time devices. Commercial availability: CP – pulsed; CCW – c.w.

wave monitoring or measuring system. Only time will be the judge of the ultimate value of these devices. A summary of the 'state-of-the-art' performance in the development laboratory (single best results) and in commercial availability is given in Fig. 10.1.

Appendix I: The equal-areas rule

The following calculation is the basis of large-signal, non-linear calculations of steady-state domain properties. The equations to be satisfied are current continuity

$$J = nev(E) + \epsilon\epsilon_0 \frac{\partial E}{\partial t} - eD \frac{\partial n}{\partial x} \qquad \text{(AI. 1)}$$

and Poisson's equation

$$\epsilon\epsilon_0 \frac{\partial E}{\partial x} = (n - N_D)e. \qquad \text{(AI. 2)}$$

N_D is the doping density, n is the electron density, and D is the diffusion coefficient. The boundary conditions for a freely propagating stable domain are

$$E \to E_R \, (= \text{constant}) \text{ as } x \to \pm \infty.$$

Therefore

$$n \to N_D, \, \partial n / \partial x \to 0 \text{ and } \partial E / \partial t \to 0 \text{ as } x \to \pm \infty ,$$

and

$$J = N_D e v_R , \qquad \text{(AI. 3)}$$

where

$$v_R = v(E_R) .$$

The steady-state travelling domain has a solution of the functional form

$$E = E(x - v_D t) , \qquad \text{(AI. 4)}$$

$$n = n(x - v_D t) , \qquad \text{(AI. 5)}$$

where v_D is the domain-drift velocity. From eqn (AI. 4)

$$\frac{\partial E}{\partial t} = -v_D \frac{\partial E}{\partial x} . \qquad \text{(AI. 6)}$$

The first stage is to calculate n as a function of E; substituting eqns (AI. 3) and (AI. 6) in eqn (AI. 1) and using

$$\frac{\partial n}{\partial x} = \frac{\mathrm{d} n}{\mathrm{d} E} \frac{\partial E}{\partial x} ,$$

$$e \{N_D v_R - nv(E)\} = -\left(eD \frac{\mathrm{d} n}{\mathrm{d} E} + \epsilon\epsilon_0 \, v_D \right) \frac{\partial E}{\partial x} .$$

Appendix I: The equal-areas rule

Eliminating $\partial E/\partial x$ with eqn (AI. 2) we have

$$\{N_D v_R - n v(E)\} = -\left(eD \frac{\mathrm{d}n}{\mathrm{d}E} + \epsilon\epsilon_0 \, v_D\right) \frac{(n - N_D)}{\epsilon\epsilon_0},$$

therefore

$$(n - N_D) \frac{\mathrm{d}n}{\mathrm{d}E} = \frac{\epsilon\epsilon_0}{eD} \{n v(E) - N_D v_R - v_D(n - N_D)\}.$$

Integrating this equation from conditions at infinity to any point in the domain,

$$\int_{n=N_D}^{n=n} \mathrm{d}\left(\frac{n}{N_D} - \ln n\right) = \frac{\epsilon\epsilon_0}{eN_D D} \int_{E_R}^{E} \left\{\left(v(E) - v_D\right) - \frac{N_D}{n}\left(v_R - v_D\right)\right\} \mathrm{d}E,$$

therefore

$$\frac{n}{N_D} - 1 - \ln\left(\frac{n}{N_D}\right) = \frac{\epsilon\epsilon_0}{eN_D D} \int_{E_R}^{E} \left\{\left(v(E) - v_D\right) - \frac{N_D}{n}\left(v_R - v_D\right)\right\} \mathrm{d}E. \tag{AI. 7}$$

If eqn (AI. 7) is to describe the domain correctly, the maximum field E_p must occur between the accumulation layer and the depletion layer, so that $n = N_D$ when $E = E_p$. v_R must equal v_D otherwise the integral on the right-hand side of eqn (AI. 7) depends on whether the integral was taken along the depletion branch with $n < N_D$ or along the accumulation branch with $n > N_D$. This would imply that $n \neq N_D$ for $E = E_p$, which is incorrect. Therefore

$$v_R = v_D, \tag{AI. 8}$$

so that the domain drifts at the same velocity as the electron-drift velocity outside the domain. Putting $v_R = v_D$ in eqn (AI. 7),

$$\frac{n}{N_D} - 1 - \ln\left(\frac{n}{N_D}\right) = \frac{\epsilon\epsilon_0}{eN_D D} \int_{E_R}^{E} \{v(E) - v_R\} \mathrm{d}E. \tag{AI. 9}$$

At the peak field of the domain $E = E_p$ and $n = N$, and eqn (AI. 9) reduces to the equal-areas rule,

$$0 = \int_{E_R}^{E_p} \{v(E) - v_R\} \mathrm{d}E. \tag{AI.10}$$

The domain shape may be calculated by integration in eqn (AI.9). An analytical approximation is usually used for $v(E)$ in order to allow analytical evaluation of eqn (AI. 9).

Appendix II: Relationship between the gain and bandwidth of a simple negative-conductance reflection amplifier

The negative conductance G and capacity C of the device are assumed to be frequency-independent at all frequencies of interest and are mounted at the end of a transmission line with a characteristic admittance Y_0 as illustrated in Fig. AII. 1.

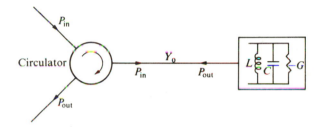

Fig. AII. 1.

The susceptance of C is often appreciably greater than the conductance of G, so it becomes necessary to put the inductance L in parallel with C and G in order to open-circuit the susceptance by resonance at the desired centre frequency so that all the r.f. current flows through G.

Defining ω_0 by $\omega_0^2 LC = 1$, the terminating admittance is

$$Y(\omega) = -G + j\omega C + \frac{1}{j\omega L} = -G + j\omega C \left(1 - \frac{\omega_0^2}{\omega^2}\right).$$

The power reflection coefficient is (Montgomery *et al.* 1948)

$$g(\omega) = \left| \frac{Y_0 - Y(\omega)}{Y_0 + Y(\omega)} \right|^2$$

$$= \frac{(Y_0 + G)^2 + \omega^2 C^2 \{1 - (\omega_0/\omega)^2\}^2}{(Y_0 - G)^2 + \omega^2 C^2 \{1 - (\omega_0/\omega)^2\}^2}. \qquad \text{(AII. 1)}$$

The maximum gain occurs when $\omega = \omega_0$ and is given by

$$g_c = \frac{(Y_0 + G)^2}{(Y_0 - G)^2}. \qquad \text{(AII. 2)}$$

119

Appendix II:

The 3 dB points occur at ω_1 and ω_2, where

$$g(\omega_{1,2}) = \tfrac{1}{2}g_c .$$

ω_1 and ω_2 are given by

$$\frac{(Y_0 + G)^2 + \omega^2 C^2 \{1 - (\omega_0/\omega)^2\}^2}{(Y_0 - G)^2 + \omega^2 C^2 \{1 - (\omega_0/\omega)^2\}^2} = \frac{1}{2}\frac{(Y_0 + G)^2}{(Y_0 - G)^2} ,$$

which reduces to

$$\omega C \left\{1 - \left(\frac{\omega_0}{\omega}\right)^2\right\} = \pm \frac{(Y_0 - G)}{\{1 - (2/g_c)\}^{\frac{1}{2}}} \qquad \text{(AII. 3)}$$

$+$ and $-$ refer to the upper and lower 3 dB frequencies ω_1 and ω_2. The bandwidth B is given by $\omega_1 - \omega_2 = 2\pi B$.

From eqn (AII. 3),

$$\omega_1 C \left\{1 - \left(\frac{\omega_0}{\omega}\right)^2\right\} = -\omega_2 C \left\{1 - \left(\frac{\omega_0}{\omega_2}\right)^2\right\} ,$$

which reduces to

$$\omega_1 \omega_2 = \omega_0^2 . \qquad \text{(AII. 4)}$$

Therefore

$$2\pi B = \omega_1 - \omega_2 = \omega_1 \left\{1 - \left(\frac{\omega_0}{\omega_1}\right)^2\right\} . \qquad \text{(AII. 5)}$$

From eqns (AII. 3) and (AII. 5),

$$2\pi B = \frac{1}{C}\frac{(Y_0 - G)}{\{1 - (2/g_c)\}^{\frac{1}{2}}} . \qquad \text{(AII. 6)}$$

From eqn (AII. 2),

$$g_c^{\frac{1}{2}} = 1 + \frac{2G}{(Y_0 - G)} . \qquad \text{(AII. 7)}$$

Combining eqns (AII. 6) and AII. 7),

$$(g_c^{\frac{1}{2}} - 1) \left(1 - \frac{2}{g_c}\right)^{\frac{1}{2}} B = \frac{G}{\pi C} , \qquad \text{(AII.8)}$$

which simplifies to

$$g_c^{\frac{1}{2}} B = G/\pi C , \qquad \text{(AII. 9)}$$

when $g_c \gg 1$. $G/\pi C$ is a frequency that can be regarded as a figure of merit for the amplifying device. It does not contain any circuit parameters.

Appendix III: Added-power and gain-saturation characteristics of reflection amplifiers

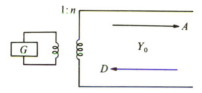

Fig. AIII. 1.

As shown in Appendix II the device admittance is purely conductive at the centre frequency. Using the equivalent circuit of Fig. AIII. 1 and the techniques outlined in § 7.3 we have

$$\frac{G}{n^2 Y_0} = \frac{A^2 - D^2}{A^2 + D^2 + 2AD\cos\theta}, \tag{AIII. 1}$$

$$n|V_n| = (A^2 + D^2 + 2AD\cos\theta)^{\frac{1}{2}}, \tag{AIII. 2}$$

where V_n is the r.f. voltage amplitude, at the diode terminals, which defines the onset of non-linearity. A and D are real voltage amplitudes of the outgoing and incoming signals, respectively, at $x = 0$ and θ is the phase difference between them. The gain (and by relation, the bandwidth) is controlled by adjusting the transformer turns ratio n. The added power P_{net} is given by

$$P_{net} = \tfrac{1}{2} Y_0 (A^2 - D^2). \tag{AIII. 3}$$

From eqns (AIII. 1), (AIII. 2), and (AIII. 3),

$$P_{net} = \tfrac{1}{2} G V_n^2. \tag{AIII. 4}$$

P_{net} is clearly independent of the amplifier gain and is determined only by the device properties.

Eqn (AIII. 3) yields

$$\frac{A}{D} - 1 = P_{net} / \tfrac{1}{2} D^2 Y_0 = \frac{P_{net}}{P_{in}}.$$

$(A/D)^2$ is the power gain g_c so we have

Appendix III:

$$g_c - 1 = P_{net}/P_{in} .$$

For convenience, the r.f. voltage to generate appreciable non-linearity is defined by the condition that P_{net} is a maximum (this condition is easy to identify experimentally). Writing $P_{in} = (P_{in})_n$ and $g_c = g_{cn}$,

$$g_{cn} - 1 = (P_{net})_{max}/(P_{in})_n . \qquad \text{(AIII. 5)}$$

Introductory references

BOWERS, R. (1966). A solid-state source of microwaves. *Scient. Am.* **215**, 21—31.

ENGELBRECHT, R.S. (1967). Bulk-effect devices for future transmission systems. *Bell Lab. Rec.* **45**, 192—8.

HILSUM, C. (1967). Miniaturizing radar. *Sci. J.* **3**, 74—9.

GUNN, J.B., BRUNQUELL, C., DAVIES, C., and REEM, G. (1967). Solid-state microwave devices: Domain and L.S.A. modes. *Electro-technology*, , 67—80.

HEEKS, J.S., KING, G., and SANDBANK, C.P. (1968). Transferred-electron bulk effects in gallium arsenide. *Elect. Commun.* **43**, 334—45.

COULTAS, F.W. (1972). Non-contacting measurements using mini-radars. *Chart. Mech. Eng.* **19**, 62—6.

SOBOL, H., and STERZER, F. (1972). Microwave power sources, *IEEE Spectrum.* **8**, 20—33.

Special-issue journals and books

IEEE Trans. Electron Devices **ED-13** (1966).
IEEE Trans. Electron Devices **ED-14** (1967).
IEEE Trans. Electron Devices **ED** 17 (1970).
IEEE Trans. Microwave Theory Tech. **MTT-18** (1970).
Proc. IEEE **59** (1971).
BULMAN, P.J., HOBSON, G.S., and TAYLOR, B.C. (1972). *Transferred-electron Devices.* Academic Press, New York.
CARROLL, J.E. (1970). *Hot electron microwave generators.* Arnold, London.

References

AITCHISON, C.S., CORBEY. C.D., and NEWTON, B.H. (1969). *Electron. Lett.* **5**, 36−7.

ALBRECHT, P., and BECHTELER, M. (1970). *Electron. Lett.* **6**, 321−2.

ALLEN, J.W., SHYAM, M., and PEARSON, G.L. (1966). *Appl. Phys. Lett.* **9**, 39−41.

BASKARAN, S., and ROBSON, P.N. (1972*a*). *Electron. Lett.* **8**, 109−10.

− − (1972*b*). *Electron. Lett.* **8**, 137−8.

BASTERFIELD, J., BOARD, K., and JOSH, M.J. (1972). Paper Fl. 6, ESSDERC. Institute of Physics, London.

BESTWICK, P., DRINAN, P., HOBSON, G.S., ROBSON, P.N., THOMAS, M., and TOZER, R. (1973). *IEEE.J. Solid State Circuits.* **SC-8**, 37−43.

BIRD, J., BOLTON, R.M.G., EDRIDGE, A.L., De Sa, B.A.E., and HOBSON, G.S. (1971). *Electron. Lett.* **7**, 300−1.

BOERS, P.M. (1973). *Electron. Lett.* **9**, 134−5.

BOLTON, R.M.G., and JONES, B.F. (1969). *Electron. Lett.* **5**, 662.

BOTT, I.B., and FAWCETT, W. (1968*a*). *Electron. Lett.* **4**, 207−9.

− − (1968*b*). *Adv. Microwaves.* **3**, 223−301.

BRADDOCK, P.W., and GRAY, K.W. (1973). *Electron. Lett.* **9**, 36−7.

BULMAN, P.J., HOBSON, G.S., and TAYLOR, B.C. (1972). *Transferred electron devices.* Academic Press, London and New York.

BUTCHER, P.N. (1965). *Phys. Lett.* **19**, 546−7.

− − (1967). *Rep. Prog. Phys.* **30**, 97−148.

− − FAWCETT, W. (1966). *Br. J. Appl. Phys.* **17**, 1425−32.

− − HILSUM, C. (1966). *Br. J. appl. Phys.* **17**, 841−50.

− − OGG, N.R. (1967). *Br. J. appl. Phys.* **18**, 755−9.

CAMP, W.O., EASTMAN, L.F., BRAVMAN, J.S., and WOODARD, D.W. (1972). *Microwaves.* **11**, 44−5.

CARSLAW, H.S., and JAEGER, J.B. (1959). *Conduction of heat in solids* (2nd edn), Oxford University Press.

CAWSEY, D. (1967). *Electron. Lett.* **3**, 550−1.

− − (1970). *IEEE.J. Solid State Circuits.* **SC-5**, 82−4.

CHARLTON, R.W., FREEMAN, K.R., and HOBSON, G.S. (1971). *Electron. Lett.* **7**, 575−6.

− − HOBSON, G.S., and MARTIN, B. (1972). *Solid-State Electron.* **15**, 517−21.

− − (1973). *IEEE Trans. Electron Devices.* **ED-20**, 812−17.

CHEUNG, P.S., and HEARN, C.J. (1972). *Electron. Lett.* **8**, 79−80.

References

CLEMETSON, W.J., KENYON, N.D., KUROKAWA, K., OWEN, B., W.D. and SCHLOSSER, W.D. (1971). *Bell Syst. tech. J.* **50**, 2917–45.

COLLIVER, D.J., and PREW, B. (1972). *Electronics.* **45**, 110–13.

CONWELL, E.M. (1967). *High-field transport in semiconductors*, Academic Press, London and New York.

– – VASSELL, M.O. (1968). *Phys. Rev.* **166**, 797–821.

COPELAND, J.A. (1967*a*). *IEEE Trans. Electron Devices* **ED-14**, 55–8.

– – (1967*b*). *IEEE Trans. Electron Devices* **ED-14**, 497–500.

– – (1967*c*). *J. appl. Phys.* **38**, 3096–101.

CORBEY, C.D., and GOUGH, R. (1973). *IEEE Trans. Electron Devices.*-ans. Electron Devices.

COULTAS, F.W. (1972). *Chart. mech. Eng.* **19**, 62–6.

DEAN, R.H., and MATURESE, R.J. (1972). *Proc. IEEE* **60**, 1486–502.

De CAQUERAY, A., BLASQUEZ, G., and GRAFFEUIL, J. (1972). *Electron. Lett.* **8**, 217–18.

De Sa, B.A.E., and HOBSON, G.S. (1971). *IEEE Trans. Electron Devices* **ED-18**, 557–62.

– – (1973). *Solid-State Electron.* **16**, 1261–6.

– – KOCABIYIKOGLU, Z.U., and HOBSON, G.S. (1971). *Electron. Lett.* **7**, 551–2.

DOWNING, B.J., and MYERS, F.A. (1971). *Electron. Lett.* **7**, 407–9.

EASSON, R.M. (1971). *Microwave J.* **14**, 53–8.

EASTMAN, L.F. (1972). *Electron. Lett.* **8**, 149–51.

EDWARDS, D., KELLETT, G., TURNER, P., and MYERS, F. (1972). *Electron. Lett.* **8**, 596–7.

EVANS, A.G.R., ROBSON, P.N., and STUBBS, M.G. (1972). *Electron. Lett.* **8**, 195–6.

FAULKNER, E.A., and MEADE, M.L. (1968). *Electron. Lett.* **4**, 226.

FAWCETT, W., BOARDMAN, A.D., and SWAIN, S. (1970). *J. Phys. Chem. Solids* **31**, 1963–90.

– – REES, H.D. (1969). *Phys. Lett.* **29A**, 578–9.

FOYT, A.G., and McWHORTER, A.L. (1966). *IEEE Trans. Electron Devices* **ED-13**, 79–87.

FREEMAN, K.R., and HOBSON, G.S. (1972). *IEEE Trans. Electron Devices* **ED-19**, 62–71.

– – (1973). *IEEE Trans. Electron Devices.* **ED-20**, 891–903.*ces.*

GILES, T.S., and SAW, J.E. (1972). *Mullard tech. Commun.* **12**, 114–19.

GLOVER, G.H. (1971). *J. appl. Phys.* **42**, 4025–34.

GUNN, J.B. (1963). *Solid State Commun.* **1**, 88–91.

– – (1964). *IBM Jl Res. Dev.* **8**, 141–59.

– – ELLIOT, B. (1966). *Phys. Lett.* **22**, 369–71.

HARTNAGEL, H.L., and IZADPANAH, S.H. (1968). *Radio electron Engr* **36**, 247–55.

HASHIGUCHI, S., and OKOSHI, T. (1971). *IEEE Trans. Microwave Theory Tech.* **MTT-19**, 686–91.

HARRIS, J.S., NANNICHI, Y., PEARSON, G.L., and DAY, G.F. (1969). *J. appl. Phys.* **40**, 4575–81.

HILSUM, C. (1962). *Proc. Inst. Radio Engrs.* **50**, 185−9.
− − REES, H.D. (1970). *Electron. Lett.* **6**, 277−8.
− − (1972). *Electron. Lett.* **8**, 373−4.
HOBSON, G.S. (1966*a*). *Electron. Lett.* **2**, 207−9.
− − (1966*b*). *Proceedings of MOGA 1966*, pp. 314−18. IEE Conference Publication no. 27.
− − (1967). *IEEE Trans. Electron Devices* **ED-14**, 526−31.
− − (1969*a*). *J. Phy.* **D2**, 1203−13.
− − (1969*b*). *Solid-State Electron.* **12**, 711−17.
− − (1972*a*). *Solid-State Electron.* **15**, 431−41.
− − (1972*b*). *Solid-State Electron.* **15**, 1107−12.
− − KOCABIYIKOGLU, Z.U., and MARTIN, B. (1970). *Proceedings of MOGA 1970*, pp. 6. 1−6.6. Kluwer-Deventer, Amsterdam.
HOLSTROM, D. (1967). *IEEE Trans. Electron Devices* **ED-14**, 464−9.
HUANG, H.G., and MACKENZIE, L.A. (1968). *Proc. IEEE* **56**, 1232−3.
HUTSON, A.R., JAYARAMAN, A., CHYNOWETH, A.G., CORIELL, A.S., and FELDMAN, W.L. (1965). *Phys. Rev. Lett.* **14**, 639−41.
ITO, Y., KOMIZO, H., and SASAGAWA, G. (1970). *IEEE Trans. Microwave Theory Tech.* **MTT-18**, 890−7.
JEPPESEN, P., and JEPPSSON, B.J. (1971). *IEEE Trans. Electron Devices* **ED-18**, 439−49.
− − (1972). *Proc. IEEE* **60**, 452−4.
JONES, D., and REES, H.D. (1972). *Electron. Lett.* **8**, 363−4.
KANBE, H., SHIMIZU, N., and KUMABE, K. (1973). *Electron. Lett.* **9**, 29−30.
KATAOKA, S., TATENO, H., and KAWASHIMA, M. (1969). *Electron. Lett.* **5**, 48−50.
KENNEDY, W.K. (1966). *Proceedings of MOGA 1966*, pp. 268−72. IEE Conference Publication No. 27.
KINO, G.S., and ROBSON, P.N. (1968). *Proc. IEEE* **56**, 2056−7.
KOCABIYIKOGLU, Z.U., and HOBSON, G.S. (1973). *Int. J. Electron.* **35**, 33−47. *J. Electron.*
KROEMER, H. (1964). *Proc. IEEE* **52**, 1736.
− − (1966). *IEEE Trans. Electron Devices* **ED-13**, 27−40.
− − (1967). *IEEE Trans. Electron Devices* **ED-14**, 476−92.
KUNO, H.J. (1969). *Electron. Lett.* **5**, 232.
KUROKAWA, K. (1967). *Bell Syst. tech. J.* **46**, 2235−59.
LAZARUS, M.J., BULLIMORE, E.D., and NOVAK, S. (1971). *Proc. IEEE* **59**, 812−14.
L.E.P. (1972). *Microwave J.* **15**, 58H.
McCUMBER, D.E., and CHYNOWETH, A.G. (1966). *IEEE Trans. Electron Devices* **ED-13**, 4−21.
McKELVEY, J.P. (1966). *Solid state and semiconductor Physics*. Harper and Row, New York.
MAGARSHAK, J., and MIRCEA, A.E. (1970). *Proceedings MOGA 1970*, pp. 16. 19−16.23. Kluwer-Deventer, Amsterdam.
MAHROUS, S., and ROBSON, P.N. (1966). *Electron. Lett.* **2**, 107−8.
M.E.S.L. (1972). *Microwave J.* **15**, 68B−68D.

References

MONTGOMERY, C.S. DICKE, R.H., and PURCELL, E.M. (1948). *Principles of microwave circuits.* Radiation Laboratory Series No. 8, McGraw-Hill, New York.

OMORI, M. (1969). *Proc. IEEE* **57**, 97.

OWENS, R.P., and CAWSEY, D. (1970). *IEEE Trans. Microwave Theory Tech.* **MTT-18**, 790–8.

PAIGE, E.G.S. (1964). *Proc. Semicond.* **8**,

PERLMAN, B.S., UPADHYAYULA, C.O., and SIEKANOWICZ, W.W. (1971) *Proc. IEEE* **59**, 1229–38.

REES, H.D. (1969). *J. Phys. Chem. Solids* **30**, 643–55.

RIDLEY, B.K. (1966). *IEEE Trans. Electron Devices* **ED-13**, 41–3.

– – WATKINS, T.B. (1961). *Proc. Phys. Soc.* **78**, 293–304.

ROBSON, P.N., and HASHIZUME, N. (1970). *Electron. Lett.* **6**, 120–1.

– – KINO, G.S., and FAY, B. (1967). *IEEE Trans. Electron Devices* **ED-14**, 612–15.

RUCH, J.G., and FAWCETT, W. (1970). *J. Appl. Phys.* **41**, 3843–50.

– – and KINO, G.S. (1967). *Appl. Phys. Lett.* **10**, 40.

– – (1968). *Phys. Rev.* **174**, 923–31.

SHOCKLEY, W. (1954). *Bell Syst. tech. J.* **33**, 799–826.

SHOJI, M. (1967). *IEEE Trans. Electron Devices* **ED-14**, 535–46.

SOLOMON, P.R., SHAW, M.P., and GRUBIN, H.L. (1972). *J. Appl.Phys.* **43**, 159–72.

STERZER, F. (1969). *Proc. IEEE* **57**, 1781–3.

SZEKELY, S., and TARNAY, K. (1968). *Electron. Lett.* **4**, 592–4.

TAYLOR, B.C., FRAY, S.J., and GIBBS, S.E. (1970). *IEEE Trans. Microwave Theory Tech.* **MTT-18**, 799–808.

THIM, H.W. (1966). *Electron. Lett.* **2**, 403.

– (1967). *IEEE Trans. Electron Devices* **ED-14**, 517–22.

– (1971). *Electron. Lett.* **7**, 106–8.

– BARBER, M.R. (1966). *IEEE Trans. Electron Devices* **ED-13**, 110–14.

WALSH, T.E., PERLMAN, B.S., and ENSTROM, R.E. (1969). *IEEE J. Solid State Circuits* **SC-4**, 374–6.

WARNER, F.L. (1966). *Electron' Lett.* **2**, 260–1.

– HERMAN, P. (1967). *Proceedings of the Cornell Conference on* high frequency generation and amplification, Cornell University, pp. 206–17.

– HOBSON, G.S. (1970). *Microwave J.* **13**, 46–52.

WASSE, M., MUN, J., and HEEKS, J.S. (1972). *Electron. Lett.* **8**, 364–6.

WILSON, K. (1969). *Mullard techn. Commun.* **10**, 286–93.

– TEBBY, A.J., and LANGDON, D. (1971). *Proceedings of 1971 European microwave Conference* Paper A6/2, pp. 1–4. Royal Swedish Academy of Engineering Sciences, Stockholm.

WOOD, C.E.C., TREE, R.J., and PAXMAN, D.H. (1972). *Electron. Lett.* **8**, 171–2.

Yu, S.P., TANTRAPORN, W., and YOUNG, J.D. (1971). *IEEE Trans. Electron Devices* **ED-18**, 88–93.

Index